# Combating Misinformation in the Age of Generative AI

The rapid advancement of Generative AI (GenAI) is revolutionizing and transforming a wide range of fields, with a notable impact on digital media and information dissemination. Digital platforms, including social media networks, have already grappled with the widespread proliferation of fake news at an increasing intensity and are now facing an even bigger threat from machine-generated and machine-boosted fake news. With the increasing growth and sophistication of GenAI models for malicious ends, fake news is spreading rapidly, finding a fertile ground in digital media for largely unchecked dissemination of incorrect and misleading information. The scalability and diversity of current and emerging GenAI techniques present both opportunities and challenges in the field of fake news. On the one hand, the ability to create highly realistic and persuasive content in large quantities with minimal effort, in real-time, makes GenAI a potent tool for spreading fake news and misinformation. On the other hand, those same capabilities can be harnessed to employ GenAI models to detect fake news and, therefore, hinder the spread. In this context, this book examines how GenAI techniques are contributing to combat fake news in digital media. A particular major challenge in combating fake news with GenAI is the explainability of the models: why a specific piece of news is determined to be fake or genuine. Tracing the origins of a specific piece of generated content—whether it stems from accurate or misleading data—requires extracting, interpreting, and representing in a legible manner vast parameter spaces and training data sources. This complexity of GenAI explainability makes it difficult to hold creators of fake news accountable and to ensure transparency in the content creation process. The authors focus on the positive potential of GenAI and provide an overview of how it can be leveraged to combat fake news across diverse digital media platforms.

The book is structured around the following key topics:

- An overview of GenAI and its relationship with fake news
- How GenAI can be utilized to fight the spread of fake news

- Ensuring that the news we encounter on digital media platforms is not misleading or false
- Methods for verifying the authenticity of news sources and tracing the origins of information to ensure its credibility
- Practical insights into technological advancements and solutions related to the detection and analysis of fake news

Finally, this book serves as a platform for the community to share innovative research on the use of GenAI in detecting and mitigating the impact of false information across digital platforms. Readers will explore how GenAI can revolutionize fact-checking, automate the identification of misleading content, and provide adaptive solutions to prevent the spread of fake news.

# Combating Misinformation in the Age of Generative AI

## Current Trends and Future Challenges

Edited by

Mohamed Lahby, Satu Elisa Schaeffer, Yassine Maleh, and Jyoti Sekhar Banerjee

CRC Press
Taylor & Fran is Group
Boca Raton London New York

CRC Press is an imprint of the
Taylor & Francis Group, an **informa** business

Designed cover image: Shutterstock ID: 2572510421

First edition published 2027
by CRC Press
2385 NW Executive Center Drive, Suite 320, Boca Raton FL 33431

and by CRC Press
4 Park Square, Milton Park, Abingdon, Oxon, OX14 4RN

*CRC Press is an imprint of Taylor & Francis Group, LLC*



ISBN: 9781041040408 (hbk)
ISBN: 9781041040415 (pbk)
ISBN: 9781003626558 (ebk)

DOI: 10.1201/9781003626558

Typeset in Sabon
by codeMantra

# Contents

# Preface

The rapid advancement of generative artificial intelligence (GenAI) is profoundly transforming numerous domains, with a particularly significant impact on digital media and information dissemination. Digital media platforms, especially social networks, which have long struggled with the proliferation of fake news, are now facing an even greater threat from machine-generated and machine-amplified misinformation. As GenAI models become increasingly powerful and accessible, malicious actors are leveraging them to produce and spread deceptive content at an unprecedented scale and speed, exploiting the openness of digital ecosystems for unchecked dissemination.

The scalability and sophistication of current and emerging GenAI models present both formidable challenges and promising opportunities. On the one hand, the capacity to create vast amounts of realistic and persuasive synthetic content in real time makes GenAI a potent tool for spreading misinformation. On the other hand, these same capabilities can be harnessed to detect, analyze, and counter the spread of fake news. This duality lies at the heart of the present volume, which examines how GenAI technologies can be used not only to understand but also to combat fake news across digital platforms.

This book presents the latest cutting-edge research, theoretical advancements, and innovative applications of GenAI in combating fake news and misinformation across digital media platforms. It examines how modern AI-driven frameworks, including large language models and hybrid AI systems, can be effectively applied to detect, analyze, and mitigate the spread of falsified or misleading information. Ultimately, this book aspires to serve as a comprehensive reference for researchers, practitioners, policymakers, and educators seeking to understand and leverage GenAI technologies responsibly in the global effort to protect truth in the digital age.

This book contains a total of 13 chapters classified into four main parts. Part I (Chapters 1 and 2) presents a state of the art related to the use of GenAI techniques to combat misinformation. Part II (Chapters 3–6) explores the transformative role of LLMs in misinformation detection. Part III (Chapters 7–10) presents case studies and practical applications from various countries, all focused on misinformation and GenAI. Finally, the last part of the book (Chapters 11–13) addresses the moral, legal, and societal implications of using GenAI in the fight against misinformation, as well as future directions for research and practice.

Finally, we want to take this opportunity to express our sincere thanks to the contributors to this volume and the reviewers for their outstanding efforts in reviewing and providing interesting feedback to the authors of the chapters. The editors would like to thank Ms. Mansi Kabra for the editorial assistance and support to produce this important scientific work. Without this collective effort, this book would not have been possible to be completed.

# About the Editors

**Mohamed Lahby** is an Associate Professor at the Higher Normal School (ENS), University Hassan II of Casablanca, Morocco. He is also the Director of the Mathematics, Artificial Intelligence, and Digital Learning Laboratory (MIND-LAB). Since 2022, he has been recognized as a senior member of the IEEE. He obtained his PhD in Computer Science in 2013 from the Faculty of Sciences and Technology of Mohammedia, University Hassan II of Casablanca. His research interests include wireless communication and networking, mobility management, QoS/QoE, the Internet of Things, smart cities, optimization, and machine learning. He has authored or co-authored more than 80 publications and has edited 15 books. He has actively participated in numerous international conferences and serves as a reviewer for several journals, including *Ad Hoc Networks*, *Applied Computing and Informatics*, and the *International Journal of Disaster Risk Reduction*. He has also chaired several international workshops and special sessions, such as MLNGSN'19, CSPSC'19, MLNGSN'20, MLNGSN'21, AI2SC'20, WCTCP'20, CIOT'22, ISGTA'23, and ISGAIE'2025.

**Satu Elisa Schaeffer** is Dean of Technology at LaSalle College in Montreal, Canada. After finishing their doctorate at Helsinki University of Technology (now part of Aalto University), Dr. Schaeffer was Professor at the School of Mechanical and Electrical Engineering at UANL in Monterrey, Mexico, and then Associate Professor of Applied Digital Intelligence at the School of Continuing Studies of McGill University in Canada. They hold a keen interest in computational intelligence, and their research explores potential applications of graph theory and machine learning on topics from areas such as social sciences, medicine, economics, and forestry. Dr. Schaeffer is passionate about digital learning, real time and asynchronous, and an early adopter of emerging software and hardware solutions to facilitate teaching and research, especially open-source solutions.

**Yassine Maleh** is an Associate Professor of Cybersecurity and IT Governance at Sultan Moulay Slimane University, Morocco. He is the founding chair of the IEEE Consultant Network, Morocco, and the founding president of the African Research Center of Information Technology & Cybersecurity. He is a senior member of IEEE and a member of the International Association of Engineers (IAENG) and the Machine Intelligence Research Labs. He has made contributions in the fields of information security and privacy, Internet of Things security, and wireless and constrained networks security. His research interests include information security and privacy, Internet of Things, network security, information system, and IT governance. He has published over 140 papers (book chapters, international journals, and conferences/workshops), 30 edited books, and five authored books. He is the editor-in-chief of the *International Journal of Information Security and Privacy (IJISP)* and the *International Journal of Smart Security Technologies (IJSST)*. He serves as an associate editor for *IEEE Access* (2019 Impact Factor 4.098), the *International Journal of Digital Crime and Forensics (IJDCF)*, and the *International Journal of Information Security and Privacy (IJISP)*. He is a series editor of *Advances in Cybersecurity Management* by CRC Taylor & Francis. He was also a guest editor of many special issues in *IEEE Transactions on Industrial Informatics*, *IEEE Engineering Management Review*, *Big Data*, *Sensors*, etc. He has served and continues to serve on executive and technical program committees and as a reviewer of numerous international conferences and journals such as *Elsevier Ad Hoc Networks*, *IEEE Network Magazine*, *IEEE Sensors Journal*, *ICT Express*, and *Springer Cluster Computing*. He was the Publicity Chair of BCCA 2019 and the General Chair of the MLBDACP 19 symposium and ICI2C'21 Conference. He received the Publons Top 1% reviewer award for 2018 and 2019.

**Jyoti Sekhar Banerjee** is currently serving as the Associate Professor and Head of the Department in the Computer Science and Engineering (AI & ML) Department at the Techno Bengal Institute of Technology (formerly known as Bengal Institute of Technology), Kolkata, India. He is also the Professor-in-Charge, R & D and Consultancy Cell & Nodal Officer of the IPR Cell of BIT. Since 2024, he also works as a Remote Researcher in the Internet of THings & AppliCAtions Lab (ITHACA) at the Department of Electrical and Computer Engineering, University of Western Macedonia, Greece. Dr. Banerjee is also working as the Adjunct Research Faculty under the Lincoln Global Postdoctoral Researcher (LGPR) Programme at Lincoln University College, Malaysia. He is the former Remote Research Fellow of the Cognitive Computing and Brain Informatics Research Group (CCBI) at Nottingham Trent University (NTU), UK. Dr. Banerjee did his Post-Doctoral Fellowship at Nottingham Trent University, UK, in the Department of Computer Science. He also completed the Post Graduate Diploma in IPR & TBM from MAKAUT, WB. He has teaching and research experience spanning 21 years and completed one IEI funded project. He is the present Secretary-cum-Treasurer of the ISTE WB Section and Secretary of the IETE, Kolkata Centre. He is the Immediate Past Secretary of the Computer Society of India, Kolkata Chapter. Dr. Banerjee is also the Vice Chairman Cum Chairman Elect in Computer Society of India, Kolkata Chapter for the year 2025–2027.

# Contributors

**Meysam Abedi**
School of Computing
University of Eastern Finland
Joensuu, Finland

**Atri Adhikari**
Independent Researcher
Kolkata, West Bengal, India

**Nahid Alaamri**
Faculty of Legal, Economic and Social Sciences
Cadi Ayyad University
Marrakech, Morocco

**Qasem Abu Al-Haija**
Cybersecurity Department
Jordan University of Science and Technology
Irbid, Jordan

**Shahed Al-Tamimi**
Cybersecurity Department
Princess Sumaya University for Technology
Amman, Jordan

**Bouchra Ayad**
Faculty of Legal, Economic and Social Sciences
Hassan II University of Casablanca
Casablanca, Morocco

**Marina Bagić Babac**
Faculty of Electrical Engineering and Computing
University of Zagreb
Zagreb, Croatia

**Ali Bouabid**
Institute of Education Sciences (ISE)
Mohammed VI Polytechnic University
Rabat, Morocco

**Kakali Das**
Greater Kolkata College of Engineering and Management
Baruipur, West Bengal, India

**Pijush Dutta**
Greater Kolkata College of Engineering and Management
Baruipur, West Bengal, India

**Abir El Aidi**
Faculty of Legal, Economic and Social Sciences
Cadi Ayyad University
Marrakech, Morocco

**Hajar El-Mahdad**
ECAF Lab FSJESAC
Hassan II University of Casablanca
Casablanca, Morocco

**Sanja Hanić**
Faculty of Electrical Engineering and Computing
University of Zagreb
Zagreb, Croatia

**Renuka Jaggi**
Himachal Pradesh National Law University
Shimla, Himachal Pradesh, India

**Samia Jirari**
Faculty of Legal, Economic and Social Sciences
Cadi Ayyad University
Marrakech, Morocco

**Mustapha Khiati**
FSBM & ECAF Lab FSJESAC
Hassan II University of Casablanca
Casablanca, Morocco

**A. V. Senthil Kumar**
Department of Computer Applications
Hindustan College of Arts & Science
Coimbatore, Tamil Nadu, India

**Aziz Mimoudi**
Institute of Education Sciences (ISE)
Mohammed VI Polytechnic University
Rabat, Morocco

**Amrita Mitra**
Greater Kolkata College of Engineering and Management
Baruipur, West Bengal, India

**Kouider Mokhtari**
Department of Teacher Education
The University of Texas
Tyler, Texas

**Vedran Mornar**
VERN University
Zagreb, Croatia

**Nouhayla Moutadid**
LBS Lab Morocco
Hassan II University of Casablanca
Casablanca, Morocco

**Cristóbal Fernández Muñoz**
Department of Communication Theories and Analysis Complutense
University of Madrid
Madrid, Spain

**Van Chien Nguyen**
School of Economics and Finance
Thu Dau Mot University
Ho Chi Minh City, Vietnam

**Revati Ramrao Rautrao**
Dr D. Y. Patil B-School
Pune, Maharashtra, India

**T. S. Sanjayan**
Department in Education
GVM's Dr Dada Vaidya College of Education
Goa University
Panaji, Goa, India

**Ismaila Temitayo Sanusi**
School of Computing
University of Eastern Finland
Joensuu, Finland

**Jasmeet Singh Sobti**
Department of Electronics & Communication Engineering
Rayat Bahra University
Mohali, Punjab, India

**Markku Tukiainen**
School of Computing
University of Eastern Finland
Kuopio, Finland

**Sonia Vatta**
Department of Computer Science & Engineering
Rayat Bahra University
Mohali, Punjab, India

**Rahul Vedisetty**
Electrical Engineering and Computer Science
Wayne State University
Wayne, Michigan

**Showkat Ahmad Wani**
School of Law
Alliance University
Bengaluru, Karnataka, India

Part I

# Misinformation in the age of GenAI

## State of the art

Chapter 1

# Social awareness, disinformation, and reputational management in the age of artificial intelligence

*Cristóbal Fernández Muñoz*

## 1.1 INTRODUCTION

The digital revolution has gone from being a promise of shared progress in the early new century to becoming a scenario where public trust, the veracity of information, and the legitimacy of institutions are questioned almost daily in recent years.

The launch in November 2022 of ChatGPT, as a natural language system capable of generating texts indistinguishable from those written by humans, accelerated a debate already brewing in the tech sector, media, and public administration: can artificial intelligence (AI) contribute to increasing the democratic quality of our societies or, on the contrary, does it pose a threat to social stability and democracy? This question is not rhetorical. Five days after its launch, ChatGPT surpassed one million users, a figure that took Facebook ten months to reach (Altman, 2023). In just two months, it already had 100 million users. This unprecedented speed of adoption illustrates the central paradox: technology grows exponentially, while regulatory, ethical, and pedagogical safeguards advance at a slower linear pace.

The result is a governance gap that overwhelms traditional regulatory frameworks, forces a rethinking of the information industry's business models, and puts the reputation of companies and institutions at risk. The European Commission defines AI as the ability of a machine to imitate some characteristics of human intelligence, such as learning, reasoning, or creativity. It is undoubtedly something that has been worked on for a long time, but it has become a focus of attention and a real concern relatively recently.

Meanwhile, all the Western technology giants, Google, Meta, Microsoft, or Amazon, as well as large Chinese companies, are immersed in the AI race, which is seen as ushering in a new era in Internet development, similar to the leap from Web 1.0 to Web 2.0 in the mid-2000s.

This is a true revolution with all kinds of implications. Its potential impact on disinformation, privacy, and its ethical boundaries is particularly worrying. Some even speak of the demise of the human species itself, in a much less poetic version of *Blade Runner*. A group of experts and entrepreneurs

DOI: 10.1201/9781003626558-2

dedicated to developing this technology raised their voices in 2023, calling for a pause in its progress; Italy blocked ChatGPT; and the United States and China have joined forces to legislate against the risks of AI.

AI is pervasive, and today, we are particularly interested in discussing its relationship with communication, its impact on the profession, our work tools, and current and future professionals in journalism, information, communication, advertising, and public relations.

Within this framework of improvement, in 2024, we begin a new phase in this prolific collaboration, transfer, and research and development that we are developing at the Complutense University under Article 60 of the Spanish University Law, in collaboration with the communications agency "Torres y Carrera." Following several previous joint investigations, the focus over the last year has shifted to the consequences of lies—and the polarization that accompanies them—in terms of social consciousness and, by extension, their relationship with and influence on concepts such as reputation and leadership.

The research "Leadership, Reputation, and Consciousness," which will be presented in book format in Spain in 2025 (Torres et al., 2025), represents a diagnosis of a reality marked by disinformation, in a scenario redefined by technological evolution that has led to the evolution of social values and models of coexistence.

Chapter 1 initially explains the research methods and provides an overview of the connections between AI and disinformation, specifically considering the new media ecosystem, social media, Generation Z, and the regulatory framework. It then addresses the results of qualitative and quantitative research on social consciousness in Spain and the resulting reputational reconfiguration and new leadership, finally offering recommendations on public and corporate policies to address the identified challenges.

## 1.2 CONCEPTUAL AND METHODOLOGICAL FOUNDATIONS

### 1.2.1 Study objectives

This study's objectives are to analyze the relationship between AI and disinformation, specifying the technological vectors that amplify the spread of fake news.

It also aims to examine the reputational impact of disinformation on organizations, leaders, and public policies.

Based on demographic research, it explores the state of social consciousness in Spain, understood as the collective capacity to discern between facts and opinions, as well as to demand accountability.

Finally, it aims to formulate recommendations for governance, media literacy, and leadership that balance technological innovation and citizen protection.

### 1.2.2 Methodology

Source triangulation aims to maximize the study's internal and external validity (Creswell and Plano Clark, 2018). This research is based on a mixed-methodological design structured in three phases:

The first phase involves a bibliographic and regulatory review, analyzing 147 academic articles (2000–2025), 21 European directives and regulations, and 18 reports from international organizations.

Second, qualitative fieldwork was conducted with six focus groups ($n$=48), 24 in-depth interviews with opinion leaders, and two Delphi panels with professionals in communications, cybersecurity, and politics.

Finally, demographic research on institutional trust, technological risk perception, and information habits was conducted using a demographic survey representative of the Spanish population ($n$=1,287; margin of error ±2.7%; with a 95% confidence interval).

## 1.3 ARTIFICIAL INTELLIGENCE AND DISINFORMATION: A HISTORICAL SYNTHESIS

### 1.3.1 From Torres Quevedo to ChatGPT

Although the rise of AI-powered information has occurred in recent years, natural language models using AI are not new. There are multiple precedents, such as more than a century ago, specifically in 1914, when Torres Quevedo presented the "chess player," considered by many to be the first heuristic automaton capable of receiving information from the environment and making autonomous decisions (Pardo-Torres, 2019). This experiment laid the foundations for contemporary cybernetics, and even then, the question was raised at a social level: if a machine can process logical rules, why couldn't it eventually simulate human reasoning? "The machine can perform the brain work of man," headlined contemporary press articles, even stating that "the human mind would be replaced by machines."

One hundred years later, AI escaped the laboratory. The development of deep neural networks, the massive availability of data, and cloud computing capabilities have enabled the development of models like GPT-4, which excel in linguistic pattern recognition and creative generation. The phenomenon has been described as a "qualitative and quantitative" leap compared to previous computerization: AI not only calculates but also writes, argues, and persuades (Ministry of Economic Affairs and Digital Transformation, 2025).

A recent study by the UK AI Security Institute, published in August 2025 as part of a collaboration between the University of Oxford and the Massachusetts Institute of Technology, among other research centers, shows how easily AI models such as Meta's Llama 3; OpenAI's GPT-4, GPT-4.5, or GPT-4o; X AI's Grok 3; and Alibaba's Qwen can be transformed into persuasion machines (Hackenburg et al., 2025).

We are particularly interested in the angle of persuasion as the disinformation generated through AI is not naive. According to Professor Rand, a specialist researcher in information science and communication management at Cornell University and co-author of the mentioned study, "what makes these AI models persuasive is their ability to generate massive amounts of relevant evidence and communicate it effectively and understandably." The study shows that people change their minds quickly and that the effect is long-lasting. In political conversations that lasted an average of nine minutes, GPT-4o was 41% and GPT-4.5 52% more persuasive than humans. People also maintained their changes of opinion between 36% and 42% longer than when convinced by a human after a month. AI chatbots influence people's conversations more when they offer a wealth of facts and evidence to support their arguments. They are also considered 5% more persuasive when they personalize messages based on the user's age, gender, and political affiliation.

According to the authors, this "could benefit unscrupulous individuals, for example, by encouraging radical ideologies and behaviors." On the other hand, and in a much more hopeful tone, research shows that language models are capable of changing the minds of people who believe in conspiracy theories and have also been shown to reduce skepticism surrounding climate change and vaccines by up to 20%.

The study supports previous findings from the London School of Economics (Costello et al., 2024), which found that AI models are more effective than humans in changing people's minds or confusing users. The researchers note that AI's ability to influence opinions also works beyond politics and has clear implications for marketing, advertising, and public relations.

Speaking to the *Financial Times* in August 2025, Dawn Bloxwich, head of innovation and development at Google DeepMind, stated that persuasion is an active and important area of research at the company. "We believe it's critical to understand the process of how AI persuades so we can build better security devices so that AI models are helpful, not harmful." Google DeepMind employs a wide variety of techniques to detect unwanted influences, such as creating classifiers that can detect manipulative language or using advanced training techniques that reward rational communication (Heikkilä, 2025).

For its part, OpenAI has stated that it does not allow political campaigns and avoids content of this type when it perfects its models.

### 1.3.2 Gen Z and disinformation

It is especially important to address disinformation and the younger generations who will constitute the citizens of tomorrow. Young individuals aged between 16 and 24—commonly referred to as Generation Z, post-millennials, or centennials—represent the first cohort to experience seamless and sustained access to the Internet from an early age. Unlike previous generations, young people today have been continuously immersed in the digital environment, where technology and online resources play a fundamental role in both their educational development and personal lives. However, their status as digital natives does not exempt them from the challenges posed by disinformation.

Our recent study on this topic (Fernández Muñoz et al., 2024) employed a mixed-methodological approach, combining quantitative data from a national demographic survey with nearly 1,500 participants and qualitative insights derived from focus groups composed of young people from diverse educational and sociocultural backgrounds. This study aims to examine the perception of fake news in Spanish society, with a particular focus on the attitudes and experiences of Generation Z. Key objectives included identifying the media outlets most consulted by these young people, understanding the motivations underlying their decisions to share content on social media, and exploring their conceptual frameworks for engaging with disinformation.

The results revealed that Generation Z is the demographic group that is most disruptive to traditional communication paradigms. They are simultaneously the most exposed to social media and the least engaged with conventional media, displaying a notable indifference toward both traditional journalism and the broader phenomenon of disinformation.

### 1.3.3 Disinformation 3.0: distinctive features

The distinctive characteristics of disinformation today are defined by four basic elements.

On the one hand, automation, with the generation of hyper-realistic texts, images, and videos, drastically reduces the production costs of hoaxes. Lying is cheap in the digital world. On the other hand, algorithmic personalization: Platforms segment audiences and optimize dissemination based on cognitive biases and emotional patterns (European Digital Media Observatory, 2024), which improves the effectiveness of disseminating disinformation tailored to the right audience.

The speed of circulation and ubiquity of information and disinformation on social media and instant messaging allow fake news to reach millions of users in a matter of minutes, and a lack of attention to the veracity of headlines can reduce outrage and, therefore, moral condemnation (Orchinik et al., 2025).

Finally, the last distinguishing feature is opacity. Generative models act as black boxes, complicating content traceability and accountability.

## 1.4 MEDIA ECOSYSTEM AND RECONFIGURING ATTENTION

Journalism, advertising, and disinformation are increasingly involved in this scenario created by AI in recent years.

The restructuring of advertising flows largely explains the vulnerability of traditional journalism. Between 2007 and 2023, investment in search engines, websites, and social media grew by 2000%, while advertising in print, radio, and television fell by half (Arce Media, 2024). The result, in an ultra-condensed form, resulted in diminished editorial staff, less fact-checking, and a heavy reliance on click-through metrics.

Populist parties have taken advantage of this journalistic and counter-power vacuum. Campaigns based on emotionality and sensationalism—"them against us"—achieve higher social returns (shares, retweets, and comments) than institutional or deliberative messages (Fernández Muñoz et al., 2024). The democratic dilemma is evident: attention is monetized based on outrage, but permanent outrage erodes rational deliberation.

Chatbots can also be as effective as, or even more so than, advertising and can have significant effects on brand attitudes or the purchase intention of a product or service. This capability could also be a boon for companies like OpenAI and Google, which are trying to make money with their AI models by integrating advertisements into their chatbots.

## 1.5 REGULATORY FRAMEWORKS AND INSTITUTIONAL RESPONSE

AI is already the greatest concern in the centers of power; whoever leads this technology will have global primacy. The three largest regions in the world with technological capabilities are facing the challenge in different ways. In the United States, they allow private companies to compete and control the data, of course in exchange for keeping government intelligence agencies informed. In Europe, the attitude is very protective: data belong to the citizen, and great care must be taken to avoid excesses. In China, things are very simple: everything is done by and for the government, which is happy with the expectation of a Big Brother who watches over and controls the population.

### 1.5.1 European Union

The EU has opted for a safeguards-based approach, focused on the defense of fundamental rights. Three instruments articulate the strategy, as shown in Table 1.1.

*Table 1.1* EU strategy on IA regulations

| *Regulation* | *Scope* | *Main requirements* | *Year* |
|---|---|---|---|
| AI regulation | Security and rights | Risk classification, audits, and fines of up to 6% of global turnover | Pending |
| Digital Services Act | Platforms | Duty of care, management of harmful content, and algorithm transparency | 2024 |
| European Media Freedom Act | Pluralism | Creation of the European Media Services Committee; Protection against Foreign Interference | 2024 |

Commissioner Thierry Breton has defined disinformation as "a form of invasion of our digital space" that should not generate "a single euro" of profit (European Commission, 2024). This reflects the ownership staked by the Old Continent, which has pushed for Russian disinformation in recent decades.

### 1.5.2 The United States and China: geopolitical contrasts

The United States relies on self-regulation of the private sector, combined with mandatory cooperation with intelligence agencies.

China grants total control to the State: AI is a tool of governance and an extension of the surveillance apparatus (Kostka, 2023).

Both models converge on the strategic priority: whoever masters AI will dominate the next phase of the global economy. They diverge in the distribution of power between public and private actors.

## 1.6 SOCIAL AWARENESS AND POLARIZATION IN SPAIN

The word "polarization" was chosen as the term of the year 2023 by the Royal Spanish Academy, confirming its centrality in the collective imagination. Our data, based on research conducted combining different qualitative and quantitative tools, generate a worrying picture. Table 1.2 highlights some of the main conclusions of the demographic survey.

In-depth interviews provided qualitative nuances such as the following:

1. Social justice is perceived as a "contested concept," and for some, an "ideological burden."
2. There is a growing demand for "moral authorities" capable of inspiring beyond political parties.
3. Civic fatigue is evident, and also a desire to "re-politicize the public sphere" if credible channels of participation existed, which are seriously damaged.

*Table 1.2* Considerations of the Spanish population on social awareness and disinformation

| *Indicator (2025)* | *% Agreement* |
|---|---|
| "Society knows how to distinguish right from wrong" (disagree) | 80% |
| "Technology facilitates immoral behavior" (agree) | 47% |
| Groups with the lowest leadership ratings: politicians, journalists, and businesspeople | Top 3/3 |

## 1.7 REPUTATION IN THE POST-DIGITAL SOCIETY

### 1.7.1 Social reputation

Social reputation refers to the collective consideration and judgment of the communication and behavior of individuals and institutions within a community (Castaño and Arias, 2021). In the attention economy, this judgment is increasingly formed in digital environments, driven by opaque algorithms. The challenge is, therefore, twofold: on the one hand, mapping perceptions in real time and, on the other, distinguishing signals (deep trends) from mere noise (viral spikes).

### 1.7.2 CSR, ESG, and narrative coherence

The concept of Corporate Social Responsibility, formalized by Bowen (1953), has evolved into Environmental, Social, and Governance (ESG) criteria, which have become a benchmark for institutional investors. The correlation between ESG performance and reputation is strong, but not linear; mediating variables such as transparency, audience age, and prior trust modulate the effect, as several authors have noted (Castaño and Arias, 2021).

Recent cases of greenwashing demonstrate that inconsistency between rhetoric and practice is more damaging than inaction. Public scrutiny, enhanced by social media and digital activism, narrows and shortens the reputational cycle.

### 1.7.3 Organizational reputation: metrics and rankings

The communication and image function in organizations has reached such a high level of complexity in recent years that it requires the precise formalization of increasingly sophisticated management tools and techniques. Having these tools and using them regularly differentiate the effective routines of professional communication departments from professional image management. Tools such as MERCO (Villafañe, 1999) combine stakeholder surveys, financial criteria, and intangible asset audits and establish a reputation ranking against which to compare oneself with the competition. According to Corporate Excellence (2024), six out of ten communication directors in Spain consider reputation their main strategic asset, and more than half of companies have increased their budget for this area. However, the rise of AI requires adapted, evolving metrics as traditional indicators (awareness and favorability) are insufficient to capture the complexity of new online conversations.

### 1.7.4 Communication reputation and crisis management

Rayner (2002) defines reputational risk as "a potential threat to an organization's future capitalization arising from public perceptions of its current behavior." In the age of disinformation, crises multiply easily and can erupt without direct corporate action: all it takes is a simple deepfake or a coordinated bot campaign.

Organizations have never been more exposed to public scrutiny and potentially critical attacks on their reputations than they are now in the digital ecosystem (Fontrodona, 2016). Crisis management takes on a new dimension, and online reputation management plays a central role in an environment where opinions and recommendations have a fundamental impact on customers.

It is important to remember that online reputation is a reflection of the prestige or esteem of an organization, person, or brand on the Internet. It depends primarily on the audience's interactions with the organization. As the fathers of public relations have already pointed out, behavior is as important as, or even more so than, its communication (Burson, 2017), even more so in a post-digital scenario.

Reputation is not under the organization's absolute control; it is also actively "manufactured" by audiences when they converse and provide their opinions. However, we do have a certain capacity to correct any deviations that may occur from the reputational status. This is especially important on the Internet, where it is very easy and cheap to disseminate information and opinions through mechanisms such as rating and review systems, social media, blogs, or forums. This is why user experience, perceived quality, and customer service are especially important in the digital environment.

Reputation is unique. It will be influenced by the organization's offline and online activities and developments. An event or occurrence that affects the organization's reputation in the physical world will also affect it in the virtual world, and vice versa.

The literature converges on three best practices:

1. 24/7 monitoring: AI-based early warning systems to detect semantic anomalies and spikes in negative sentiment.
2. Transparent and empathetic response: admitting uncertainty, explaining the facts, and detailing corrective actions.
3. Partnerships with trusted fact-checkers and media: outsourcing data validation strengthens the organization's credibility (Vallejos, 2012).

## 1.8 CRISIS OF TRUST AND NEW LEADERSHIP MODELS

Data from the Edelman Trust Barometer 2025 point to the loss of trust among key stakeholders and the impact of disinformation on this. About 69% of the population believes their leaders "deliberately lie" (Edelman Trust Institute, 2025).

The same research shows that 62% of people believe that liberal democracy "is in danger" due to disinformation and that 58% demand that companies "take a public stand" on social issues. The reality is that we are facing a profound lack of trust, and it is worrying that citizens stop considering public actors to solve problems, looking to the private sector. It seems clear that leadership models need to be reviewed.

Our research supports a new humanistic leadership as a value proposition. Núñez & Giménez Thomsen (2023) already described the new leadership model based on four pillars:

1. Self-knowledge: recognition of strengths and limitations.
2. Social compass: vision oriented toward the common good, not just corporate profit.
3. Empathic trust: ability to delegate and listen.
4. Transcendence: legacy beyond the temporary mandate.

Both the experts consulted and the focus groups we conducted corroborate the demand for "transformational, participatory, and ethical" leaders capable of articulating an inclusive narrative and fostering intergenerational engagement.

## 1.9 ILLUSTRATIVE CASE STUDIES

Disinformation is not a direct product of AI. There are thousands of cases in history of the use of lies for a wide variety of purposes. The quantitative leap in its use has occurred with the Internet and the explosion of social media, which allow for the mass dissemination of falsehoods, and the qualitative advance comes with AI. If nothing is done about it, machines will be able to construct false realities to manipulate human beings and society.

Since its birth in the mid-2000s, the digital environment has provided new possibilities for connecting citizens to a continuous, ubiquitous, and international network. This is a context where data and mobile technologies take on new meaning, modifying preexisting communication codes and generating new individual, social, and ethical challenges.

In the digital landscape, which continues to evolve rapidly and some authors are beginning to call post-digital, we find ourselves facing a panorama characterized by information hyperinflation, with greater access to information, although not necessarily better information, and with a progressive loss of credibility among traditional information players: governments, organizations, and the media.

Digital and social platforms, in their infinite possibilities, introduce new ways of relating, obtaining information, creating and sharing content, learning and intellectual development, etc. Users have been empowered since technology placed an entire media universe (real and virtual) in the palm of our hands and just a click away. Social networks have become a priority medium for the Internet, which has quickly taken a leading role in the spaces of communication, socialization, consumption, and participatory learning.

Today's cyberconnected society finds itself in a flux of data exchanges transformed into photography, video, text, themes and trends, stories, and narratives that, both individually and collectively, generate different trends such as clicktivism and digital leisure.

However, despite the fact that the Internet has become the new agora of the 21st century, its mere use does not imply successful results or constructive associations, socially or individually.

A scenario of contradictions is consolidating, where the technological hyperconnection of the public does not necessarily imply the existence of an emotionally connected or reliably informed audience.

Before 2016, we had the utopian networks (Arab Spring), then the moderator networks (arbiters of truth), and now the anything goes networks (Musk, harassment, and insults), with decisions such as the goal of eliminating content verifiers, as Meta did at the end of 2024. In the new scenario, voluntary self-monitoring of platforms is insufficient to contain organized hate campaigns. This is one of four lessons learned, included in Table 1.3, regarding disinformation during the last years.

The 2020 dramatized documentary film "The Social Media Dilemma" featured testimony from experts and former employees of large technology companies on the ins and outs of social media, which strives to capture attention to monetize and make money from advertising, but above all, warning of its inappropriate uses, as we have seen lately, for example, by Elon Musk and X, interfering in electoral processes and shaking the most basic principles of Western democracies.

Democratic governments and companies, especially those that are publicly traded, need professional media outlets to counter disinformation

*Table 1.3* Four lessons learned from recent cases of digital disinformation

| *Case* | *Description* | *Lessons* |
|---|---|---|
| Fox News vs. Dominion (2023) | $787 million settlement for spreading election hoaxes | Financial sanctions can deter disinformation, but they require documentary evidence and lengthy judicial processes. |
| Valencia DANA Crisis (2024) | Overflow of disinformation and contradictory messages regarding emergency warnings | Collective emotion prevails over reason when official communication is late or ambiguous. |
| Document "The Social Media Dilemma" (Netflix, 2020) | How social media platforms use manipulative algorithms to create addictive user experiences for profit. | The power of insider testimony to raise awareness and pressure regulations. |
| "Not verified" by Meta (2024) | Elimination of external fact-checkers on Facebook | Voluntary self-monitoring of platforms is insufficient to contain organized hate campaigns. |

created by unknown parties, even bots, and spread through social media and other platforms with spurious interests. This has been seen in various electoral processes and in the recent crisis of regional banks in the United States, where Twitter was misused.

The $787 million payment that Fox News has had to make for knowingly spreading false claims that the vote count for the last presidential election had been manipulated is a sign of a shift in sensitivity toward large-scale lies. In Europe, the EU is passing a whole battery of regulations to protect citizens from this epidemic.

As a peripheral but clearly prevalent element in the described map, we find the totalitarian currents that are spreading throughout the world and in Spain are tinged with dictatorship nostalgia. The tragedy of the DANA in Valencia, Spain, in 2024 represents a perfect test tube for contemplating the dispute for the soul of the citizenry. The idea that "people save the people," in contrast to that of a "failed state," demonstrates the prevalence of emotion over reason. This prevalence is exacerbated by disinformation, but it wounds the very essence of our social fabric. An inert society responds broadly to immediate emotional stimuli.

## 1.10 PUBLIC POLICY AND CORPORATE MANAGEMENT RECOMMENDATIONS

### 1.10.1 Media literacy and critical thinking

Perhaps, one of the main recommendations is to integrate fact-checking skills into primary and secondary education curricula to foster critical thinking, media literacy, and resilience to misinformation from an early age.

Similarly, it would be advisable to establish formal teacher certification programs in digital literacy, developed through strategic partnerships with universities and schools, to ensure that educators are trained to guide students in managing complex digital environments.

It would be very helpful to support the creation of civic–technological innovation labs that bring together students, journalists, and software developers to collaboratively design tools, campaigns, and educational resources aimed at strengthening participation and democratic values and, consequently, combating disinformation.

### 1.10.2 Algorithmic governance

In the field of algorithmic governance, full traceability of generative AI models must be required, including mandatory logging of training datasets, model versions, and update logs, to ensure transparency, accountability, and ethical oversight throughout the whole development cycle.

Furthermore, independent audits should be mandated for large-scale recommendation systems, particularly those serving millions of individual users worldwide, to assess algorithmic fairness, content moderation practices, and potential risks to democratic discourse and user autonomy.

In addition, it would be advisable to implement a tiered sanctions similar framework based on the "who pollutes, pays" principle, whereby platforms or developers responsible for reputational damage or disinformation were fined proportionally, and the funds raised were redirected to support media pluralism, investigative journalism, and civic education initiatives.

### 1.10.3 Reputational strategy for organizations

Finally, regarding recommendations related to the reputational strategy, it would be necessary to implement more robust and widespread systemic listening practices by integrating social auditing technologies, in-depth ESG (environmental, social, and governance) analysis, and internal employee surveys to capture real-time opinions, identify emerging reputational risks, and inform strategic decision-making, fostering a two-way dialogue of maximum respect.

Furthermore, it is necessary to ensure narrative coherence across all points of contact within the organization, aligning the stated purpose, internal culture, and externally visible behaviors to generate authentic trust. Superficial or symbolic messages, what we could call "decorative intent" (Amare and Manning, 2013) undermining credibility and stakeholder engagement, should be avoided.

It would be interesting and advisable to better train organizations and conduct deepfake simulations to prepare crisis management teams for scenarios involving falsified audio or video content. These drills should improve response protocols, media verification skills, and cross-functional coordination in situations of reputational pressure.

Finally, it is worth mentioning the recommendation to enhance multi-sector intelligence-sharing partnerships; collaborating with NGOs, academic institutions, and regulatory bodies to monitor disinformation trends; exchange best practices; and coordinate rapid responses to reputational threats in real time.

## 1.11 CONCLUSIONS AND FUTURE LINES OF RESEARCH

The convergence of AI, business models based on the attention economy, and setbacks in traditional journalistic filters has created an environment in which disinformation proliferates easily. However, the same technology that fuels the problem offers tools to mitigate it: anomaly detection systems, automated verifiers, and blockchain traceability. The real challenge lies in combining innovation, regulation, and civic culture.

Reputation—social, organizational, and communicational—serves as a thermometer of democratic health and a barometer of business sustainability. Managing it no longer depends solely on communication campaigns, but on the true capacity of organizations to act with coherence, transparency, and purpose.

As expected, and as optimistic academics who want to make useful social contributions, we believe in a better world: in education, in youth, social well-being, the proper place of women in the 21st century, and all the SDGs. Although we sometimes argue in academia about the transformation of journalism and the dispossession that the practice of the profession often entails today, we value, respect, and unwaveringly defend journalism, which is essential to guaranteeing democracy. And to achieve this, we must also defend from the earliest ages and as an inalienable universal right, education, knowledge, and truth.

In the face of civic fatigue and polarization, signs of hope are emerging: greater demand for accountability, consolidation of ESG standards, and new forms of humanistic leadership. Seizing this window of opportunity requires a three-pronged equation: education, regulation, and collaboration, or, to put it another way, moving from an inert society to a conscious society.

Finally, it's necessary to point out, although methodological triangulation strengthens our findings, that some limitations remain:

1. Technological dynamics: the speed of evolution of AI models can render regulations and metrics obsolete in a matter of months.
2. Cultural bias: the survey, in-depth interviews, and focus groups are limited to Spain; extrapolations should be made with caution.
3. Availability of corporate data: the opacity of proprietary algorithms limits the accuracy of discourse analysis.

Future research should explore, among other aspects, the validity of this diagnosis in other geographical, social, and cultural spheres; the influence of AI in the field of persuasive communication and information, as well as the effectiveness of platform taxes in promoting and funding independent local journalism; the interaction between generative AI and political emotions or indicators of "cognitive resilience" in vulnerable communities; and the new digital divides in the new scenario of digital access to knowledge.

## REFERENCES

Altman, S. (2023, May). Keynote address on language models and society [transcript]. OpenAI.

Amare, N., & Manning, A. (2013). *A Unified Theory of Information Design: Visuals, Text and Ethics* (1st ed.). Routledge.

Arce Media. (2024). Advertising investment in Spain 2007–2023.

Bowen, H. R. (1953). *Social Responsibilities of the Businessman.* Harper & Row.

Burson, H. (2017). *The Business of Persuasion.* Rosetta Books.

Castaño, C., & Arias, J. (2021). Corporate social responsibility and reputation: mediating variables. *Journal of Business Ethics*, 173(2), 345–362.

Corporate Excellence. (2024). Approaching the future: trends in reputation and intangible asset management.

Costello, T., Pennycock, G., & Rand, D. G. (2024). Durably reducing conspiracy beliefs through dialogues with AI. *Science*, 385, 1814. https://doi.org/10.1126/science.adq1814

Creswell, J. W., & Plano Clark, V. L. (2018). *Designing and Conducting Mixed Methods Research* (3rd ed.). SAGE.

Dominion Voting Systems v. Fox News. (2023). 1:21-cv-00040. Delaware Superior Court.

Edelman Trust Institute. (2025). Edelman trust barometer 2025.

European Commission. (2024). Proposal for a European regulation on artificial intelligence.

European Digital Media Observatory. (2024). Annual report on disinformation in Europe.

Fernández Muñoz, C., Rubio Moraga, A. L., & Álvarez Rivas, D. (2024). Generation Z and disinformation: perceptions and practices in the digital age. *Estudios del Mensaje Periodístico*, 30(3), 517–529. https://doi.org/10.5209/emp.96511

Fontrodona, J. (2016). Transparency and trust in the digital age. *Cuadernos de Ética*, 29(1), 45–60.

Hackenburg, K., Tappin, B. M., Hewitt, L., Saunders, E., Black, S., Lin, H., ... & Summerfield, C. (2025). The levers of political persuasion with conversational AI. arXiv preprint arXiv:2507.13919.

Heikkilä, M. (August 13, 2025). The art of persuasion: how top AI chatbots can change your mind. *Financial Times.* https://www.ft.com/content/31e528b3-9800-4743-af0a-f5c3b80032d0

Kostka, G. (2023). China's approach to AI governance. *Asian Survey*, 63(1), 23–49.

Merco. (2024). Merco companies and leaders ranking 2024.

Ministry of Economic Affairs and Digital Transformation. (2025). Priorities of the Spanish presidency of the EU council.

Núñez, J., & Giménez Thomsen, P. (2023). Humanistic leadership: theory and practice. *Harvard Deusto Business Review*, 319, 34–43.

Orchinik, R., Bhui, R., & Rand, D. G. (2025). Replicability and generalizability of the repeated exposure effect on moral condemnation of fake news. *Nature Communications*, 16(1), 7206.

Pardo-Torres, M. (2019). Leonardo Torres Quevedo and pioneering cybernetics. *History and Technology*, 14(2), 87–102.

Rayner, J. (2002). *Managing Reputational Risk*. John Wiley & Sons.

Torres, X., Fernández-Muñoz, C., Brey, E., & Álvarez Rivas, D. (2025). *Liderazgo, Reputación y Conciencia*. Aula Magna McGraw Hill.

Vallejos, M. (2012). Crisis and online communication. *Comunicar*, 20(39), 201–210.

Villafañe, J. (1999). *Corporate Reputation*. Prentice Hall.

Chapter 2

# Aligning generative AI with educational values

## An ethical response to digital misinformation

*Aziz Mimoudi, Kouider Mokhtari, and Ali Bouabid*

## 2.1 INTRODUCTION

### 2.1.1 Background

The rapid rise of generative AI (GenAI), particularly large language models (LLMs) such as GPT-5, is reshaping how education is designed, delivered, and experienced. These systems can personalize learning, generate instructional materials, and support more interactive engagement (Porayska-Pomsta et al., 2024). At the same time, their use introduces profound challenges. AI-generated misinformation – whether in the form of fabricated citations, hallucinated "facts," or deepfakes – threatens the credibility of academic work and undermines the values of critical thinking and intellectual autonomy (Yan et al., 2024).

This dual nature, where AI functions simultaneously as an enabler and a disrupter, underscores the pressing need for careful and ethically grounded integration. The issue is not only technical but also ethical and pedagogical: how can educators ensure that AI strengthens rather than erodes the integrity of learning?

### 2.1.2 Methodology

This chapter adopts a review – and synthesis – methodology. It:

- Reviews existing scholarship on GenAI and misinformation, along with global frameworks such as UNESCO (2021), the European Commission (2022), Stanford's AI Literacy Framework (2025), and Canada's FASTER guidelines (2025).
- Examines empirical evidence, including case studies (e.g., OBVIA, 2024) and institutional responses such as UM6P's AI use policy.
- Develops an Ethical Alignment Model (EAM) that integrates pedagogy, governance, and technical safeguards to address misinformation.
- Illustrates strategies and practices that educators and institutions can adopt to foster resilience against AI-driven misinformation.

DOI: 10.1201/9781003626558-3

### 2.1.3 Research problem, novelty, and contribution

Although numerous ethical frameworks for AI exist, few directly confront the specific challenge of misinformation in educational contexts. Even fewer provide concrete pedagogical and institutional strategies to resist misinformation while enabling responsible AI use. This chapter's unique contribution lies in the following:

1. Introducing a typology of misinformation tailored to educational contexts (accidental, ignorant, and intentional).
2. Synthesizing global ethical frameworks into a coherent EAM.
3. Translating abstract principles into practical classroom strategies and institutional policies.

### 2.1.4 Chapter organization

The rest of Chapter 2 is organized into seven sections. Section 2.2 reviews related work on misinformation, AI ethics, and education. Section 2.3 explores the ethical risks of GenAI, including hallucinations, deepfakes, and the blurring of authorship. Section 2.4 reviews international frameworks and introduces the proposed EAM. Section 2.5 outlines teaching strategies to foster AI literacy and resilience against misinformation. Section 2.6 considers how ethics and AI literacy can be embedded in curriculum design and institutional governance. Section 2.7 concludes with reflections on future challenges and opportunities for integrating GenAI responsibly in education.

## 2.2 RELATED WORK

Research on misinformation in education predates the rise of GenAI. Earlier scholarship emphasized media literacy and digital citizenship to equip learners with skills to evaluate online content critically (Keefe, 2019). With the advent of GenAI, however, the scale and sophistication of misinformation have intensified, requiring new responses.

Scholars have documented GenAI's epistemic limitations. Bender et al. (2021) describe LLMs as "stochastic parrots" – replicating linguistic patterns without ensuring factual accuracy. Marcus and Davis (2019) warn of hallucinations, where models generate plausible but false references. Abid et al. (2021) and Lucy and Bamman (2021) show how models reproduce and amplify stereotypes, raising issues of equity and bias.

At the policy level, international bodies have begun issuing guidelines. UNESCO (2021) highlights human rights, inclusion, and sustainability. The European Commission (2022) emphasizes transparency and oversight. Stanford's AI Literacy Framework (2025) defines four literacies (functional, ethical, rhetorical, and pedagogical). Canada's Trousse pédagogique (Bruneault et al., 2024) provides classroom-ready tools, while the Government of Canada's FASTER Framework (2025) offers institutional-level principles.

Despite this progress, most frameworks treat misinformation indirectly. An integrated approach that connects typologies of misinformation, ethical frameworks, and actionable strategies is missing. This chapter addresses that gap.

## 2.3 THE ETHICAL RISKS OF MISINFORMATION IN EDUCATIONAL USE OF GENAI

GenAI introduces risks that are technical, cognitive, social, and institutional.

Misinformation does not simply manifest as factual mistakes; it challenges trust, authorship, equity, and governance. This section examines the most pressing ethical risks associated with GenAI use in education.

### 2.3.1 Overview of risks from LLMs

LLMs are often criticized as opaque "black boxes." Their lack of explainability undermines accountability and limits users' ability to verify outputs (Bender et al., 2021). They produce hallucinations – false but plausible claims or references (Marcus & Davis, 2019) – and frequently replicate social and cultural biases (Adib et al., 2021; Lucy & Bamman, 2021).

Beyond accuracy, the pedagogical danger is passivity. Students may rely on fluent but shallow outputs, weakening habits of inquiry and critical reasoning (Porayska-Pomsta et al., 2024). These risks demand safeguards that combine technical transparency, institutional oversight, and deliberate cultivation of AI literacy.

### 2.3.2 Deepfakes and falsified content

The risks extend beyond text. GenAI tools now generate convincing synthetic images, audio, and video. In education, this enables the following:

- **Academic dishonesty:** students may fabricate interviews, simulate attendance in online courses, or submit falsified multimedia assignments.
- **Reputational harm:** faculty and administrators can be impersonated in fake announcements (OBVIA, 2024).
- **Disinformation campaigns:** institutions may be targeted by coordinated synthetic attacks.

Chesney and Citron (2019) describe this as the liar's dividend: when synthetic media becomes widespread, the credibility of all content – authentic or not – declines. Recent research also highlights inequalities in awareness and vulnerability to deepfakes, showing disparities in knowledge and attitudes across demographic groups (Bitton et al., 2024). Deepfakes also disrupt scientific knowledge dissemination and public communication, further complicating trust in academic and research contexts (Doss et al., 2023).

### 2.3.3 Case examples from OBVIA (2024)

The OBVIA (2024) report provides striking illustrations of the destabilizing effects of AI-generated misinformation in educational contexts. In one case, a deepfake video impersonated a university rector and falsely announced a merger between faculties, sparking protests and confusion until it was publicly refuted. Academic journals in Québec were also targeted, receiving AI-generated articles that included fabricated authors, citations, and metadata, thereby threatening the credibility of scholarly publishing. Similarly, student government elections were disrupted by deepfaked campaign materials that falsely attributed offensive remarks to candidates, distorting democratic participation on campus. Collectively, these cases demonstrate that AI-driven misinformation undermines not only classroom integrity but also the broader institutional trust and democratic engagement essential to academic life.

### 2.3.4 Impacts on trust, source credibility, and autonomy

UNESCO (2021) emphasizes that education must safeguard human agency and autonomy, while Canada's FASTER Framework (2025) warns of the dangers of automation bias when decision-making is overly delegated to machines. The risks of GenAI misinformation in education are, therefore, multifaceted. First, repeated exposure to false or synthetic content can erode trust, diminishing students' confidence in instructional communications and in the reliability of digital learning platforms (Selwyn, 2023). Second, credibility is compromised when educators unknowingly share AI-generated misinformation, leaving them vulnerable to reputational damage and weakening the authority of academic institutions. Finally, learner autonomy is reduced when students place undue reliance on AI outputs, a phenomenon that Holmes et al. (2022) describe as cognitive outsourcing, in which critical thinking and independent judgment are gradually displaced.

Mitigating these risks requires deliberate safeguards, including human-in-the-loop design, regular feedback loops, and the systemic integration of AI literacy into curricula.

Together these measures help maintain trust, protect institutional credibility, and ensure that students retain the agency to critically evaluate the AI-generated information rather than passively accept it.

### 2.3.5 The blurring of authorship and originality

GenAI also complicates academic norms of authorship. When AI tools contribute to writing or instructional design, key questions arise: Who is the legitimate author? What counts as original? Floridi and Cowls (2019) call for transparency-by-design, ensuring disclosure of AI use. The University of Glasgow, for instance, requires students to cite AI tools as personal communications and include them in acknowledgments. Such practices highlight that ethical AI use requires disclosure, documentation, and pedagogical framing.

### 2.3.6 Legal and institutional implications

AI-driven misinformation introduces legal risks across copyright, liability, and privacy. The U.S. Copyright Office (2023) affirms that fully AI-generated works cannot be copyrighted. Deepfake impersonations raise concerns about defamation. The European Commission (2022) emphasizes that AI in education must comply with privacy and data protection laws.

Canada's FASTER Framework (2025) provides structured responses, including black-box risk management, human-centered review, and transparent metadata logging. Chen and Shu (2024) further call for life-cycle countermeasures, ranging from input filtering to post-deployment audits. Conducting misinformation research itself also raises ethical concerns related to deception, participant consent, and harm mitigation, requiring careful methodological safeguards (Passi & Vorvoreanu, 2022).

Institutions must, therefore, revise integrity policies, embed AI literacy in codes of conduct, and align practices with national and international regulations. As Osborne and Pimentel (2023) argue, learners must be trained as "competent outsiders" capable of interrogating AI-generated claims.

### 2.3.7 Typology of misinformation

Not all misinformation is the same. Following Keefe (2019), misinformation in education falls into three categories: accidental, ignorant, and intentional. Table 2.1 summarizes their definitions, typical scenarios, and institutional responses.

As mentioned in Table 2.1, these three categories converge toward a unified response layer. Each requires a tailored intervention: technical safeguards for accidental errors, literacy scaffolds for ignorant cases, and governance or legal measures for intentional disinformation.

Figure 2.1 shows how the three forms of misinformation identified in Table 2.1 require distinct but interconnected responses. The response layer integrates verification practices, literacy modules, and institutional sanctions to preserve trust and academic integrity.

*Table 2.1* Typology of misinformation in education

| *Type* | *Definition in education* | *Typical scenarios* | *Primary responses* |
|---|---|---|---|
| Accidental | Model limitations (e.g., hallucinations and outdated data). | LLM fabricates references. | Verification tools; citation literacy. |
| Ignorant | User lacks background knowledge and misinterprets outputs. | Student misreads AI summary. | Media/AI literacy; scaffolding. |
| Intentional | Deliberate manipulation (deepfakes and fabricated data). | Fake rector video; impersonation. | Institutional policy; sanctions; detection tools. |

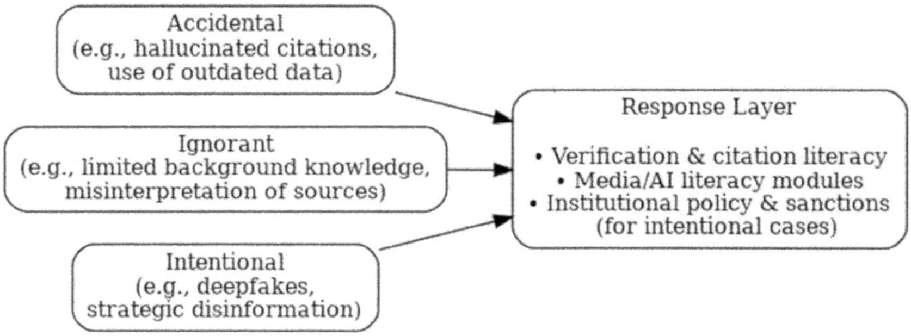


*Figure 2.1* Typology of misinformation in education and response layer.

This typology connects directly to the EAM, which builds on it by linking specific types of misinformation to pedagogical, governance, and technical responses.

## 2.4 FRAMEWORKS FOR ETHICAL ALIGNMENT OF LLMS IN EDUCATION

As GenAI becomes deeply embedded in higher education, adopting it without ethical grounding risks undermining academic integrity, fairness, and student agency. To counter these risks, particularly those associated with misinformation, a range of ethical frameworks have emerged from universities, international organizations, and policy bodies. This section reviews and critiques five influential approaches: Stanford's AI Literacy Framework, the European Commission's ethical guidelines, the Canadian Trousse pédagogique, UNESCO's Ethical Readiness Framework, and the OECD and Canada's FASTER Framework.

### 2.4.1 Stanford's AI Literacy Framework

Stanford's Teaching Commons (2025) define four dimensions of AI literacy. Functional literacy involves understanding how GenAI systems are trained, how they generate outputs, and what their technical limitations are. Ethical literacy addresses the social consequences of AI, including bias, fairness, and privacy. Rhetorical literacy examines how AI-generated content persuades or shapes perceptions – an especially relevant skill for identifying and resisting misinformation. Finally, pedagogical literacy prepares educators to integrate AI in ways that promote student agency and meaningful engagement.

The framework's strength lies in its holistic design, which extends beyond technical skills to include social and communicative dimensions. Its limitation is that it leaves institutional governance largely unaddressed. For misinformation, rhetorical literacy is particularly valuable as it helps learners analyze not only what AI claims but also how such claims are framed to influence belief.

### 2.4.2 European Commission guidelines

The European Commission (2022) published ethical guidelines on the use of AI and Data in Education, which emphasize four principles: transparency, human oversight, non-discrimination, and digital wellbeing/privacy. These guidelines are practical, focusing on teacher training and institutional safeguards. Their main strength is the establishment of clear anchors for transparency and oversight in schools and higher education.

However, the guidelines are compliance-oriented. They set out what institutions should guarantee but do less to show educators how to adapt pedagogy to counter misinformation. Without explicit classroom strategies, they risk remaining at the level of policy rhetoric.

### 2.4.3 PIA Trousse pédagogique (Canada)

The Pôle interordre de Montréal developed the Trousse pédagogique as a classroom-ready toolkit for teaching AI ethics in postsecondary education (Bruneault et al., 2024). It contains ten adaptable activities designed to present real-world dilemmas; engage students in debate on surveillance, bias, and misinformation; and balance technical fluency with moral reasoning. Operating across four dimensions – technical, moral, sociotechnical, and normative – the toolkit makes AI ethics concrete and accessible for educators.

Its strength is practicality: it allows instructors to integrate AI ethics without needing specialist expertise. Its limitation is scalability: developed in the context of Québec higher education, it requires adaptation to fit into diverse cultural and institutional environments.

### 2.4.4 UNESCO Ethical Readiness Framework

UNESCO's Recommendations on the Ethics of Artificial Intelligence (2021), adopted by 193 member states, provides a global normative framework. It emphasizes five pillars: human agency and oversight, fairness and inclusion, transparency and explainability, sustainability, and accountability and governance.

UNESCO's framework is comprehensive and unique in highlighting environmental sustainability and is an often-overlooked dimension in educational AI. Yet its limitation lies in its macro-level orientation: while it provides valuable principles, it does not outline concrete strategies for managing classroom-level challenges such as hallucinations or deepfakes.

### 2.4.5 OECD and Canada's FASTER Framework

The OECD's AI principles (2019) establish an international backbone emphasizing human-centered values, robustness, inclusiveness, and transparency. However, they remain broad and require contextual adaptation.

By contrast, Canada's FASTER Framework (2025) offers a more operational approach, organized around six principles: Fair, Accountable, Secure, Transparent, Educated, and Relevant. It provides guidance and risk assessments, black-box audits, and metadata disclosures, making it particularly useful for institutions seeking structured responses to AI risks, including misinformation. Its limitation, however, is that it was designed primarily for public administration and must be tailored to the realities of higher education.

### 2.4.6 Synthesis

Taken together, these frameworks offer complementary strengths. Stanford and the PIA toolkit emphasize classroom practice and pedagogy, the European Commission and UNESCO focus on institutional ethics and governance, and the OECD and Canada's FASTER Framework provide policy backbones and operational tools.

Yet across all of them, a notable gap remains: few explicitly address misinformation as a distinct challenge. While principles such as fairness, transparency, and accountability are emphasized, most frameworks stop short of directly confronting how GenAI hallucinates, fabricates, or amplifies disinformation in educational contexts.

To address this gap, Chapter 2 proposes an EAM that brings together insights from the reviewed frameworks and organizes them into three

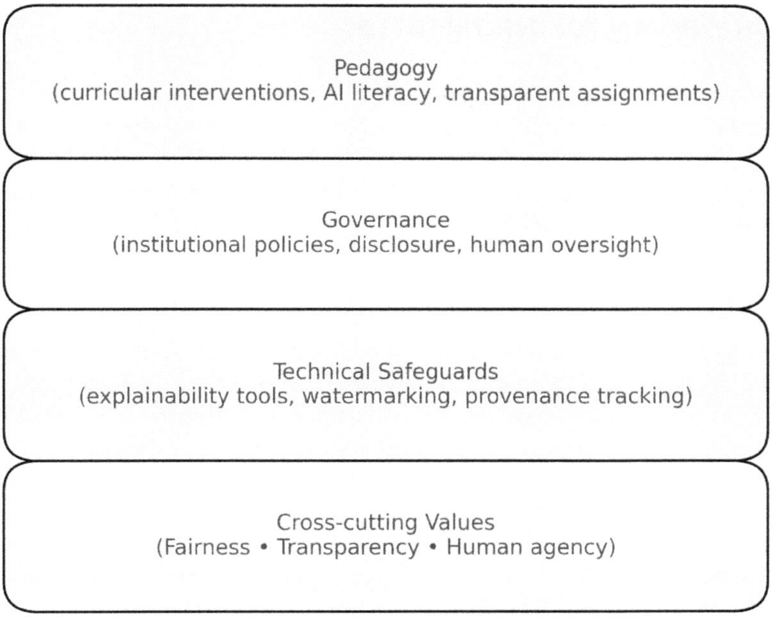


*Figure 2.2* Ethical alignment model (EAM) for GenAI and misinformation in education.

mutually reinforcing layers. The first layer, pedagogy-embedding AI literacy across curricula, focuses on cultivating rhetorical awareness, designing transparent AI-augmented assignments, and incorporating activities that train students to detect misinformation, such as spotting deepfakes and conducting fact-checks. The second layer, policy and governance, highlights the need to update academic integrity codes, strengthen disclosure requirements, carry out impact assessments, and ensure that human oversight remains central. Finally, the third layer, technical safeguards, focuses on the application of explainability tools, watermarking and provenance tracking mechanisms, access controls, and lifecycle audits, all of which contribute to greater transparency and accountability. Figure 2.2 illustrates how these three layers interact under cross-cutting values – human agency, fairness, transparency, and sustainability – to produce outcomes of trust, integrity, and critical AI literacy.

The model depicts three interconnected layers – pedagogy, governance, and technical safeguards – underpinned by cross-cutting values. Together, they form a systemic response to misinformation, ensuring that GenAI integration supports academic integrity rather than undermining it. By combining curricular interventions, institutional policies, and technical protections, the EAM enables higher education institutions to foster resilience while maintaining fairness, transparency, and human agency.

## 2.5 TEACHING STRATEGIES FOR COMBATING AI-DRIVEN MISINFORMATION

Frameworks provide guiding principles, but their effectiveness depends on how they are implemented in practice. This section translates the EAM into classroom and institutional strategies that enable students and educators to resist AI-driven misinformation while using GenAI responsibly as a learning tool.

### 2.5.1 Embedding AI literacy across disciplines

AI literacy should not remain confined to computer science or technical courses; it must be treated as a transversal competence. As Touretzky et al. (2019) argue, literacy involves not only understanding how AI works but also developing the ability to critique its outputs. Universities can embed short AI literacy modules in first-year seminars, writing-intensive courses, and civic education classes, ensuring early exposure. Assignments may require students to analyze AI-generated outputs, identify hallucinated references, and contrast them with peer-reviewed sources. Embedding AI and media literacy within STEM, humanities, and social science curricula ensures that learners across disciplines are equipped to critically evaluate GenAI before accepting its outputs as knowledge.

### 2.5.2 Teaching deepfake and synthetic media detection

As deepfakes and other synthetic media become more sophisticated, students need what can be called synthetic media literacy. Educators can address this by demonstrating common detection cues – such as unnatural blinking or mismatched lip-sync – and by integrating browser-based detection tools like InVID or Deepware Scanner into classroom exercises. Structured debates where students must analyze or defend deepfake personas can also be effective as they cultivate rhetorical awareness and reveal how persuasive manipulation operates. These activities align with both the pedagogical and technical layers of the EAM: they build students' capacity for critical engagement while exposing them to practical detection tools.

### 2.5.3 Developing fact-checking habits

Fact-checking remains a cornerstone of combating misinformation. Building on civic education models such as those developed by the Stanford History Education Group, teachers can present real-world misinformation incidents – such as fabricated institutional announcements or deepfake campaign materials – as case studies for analysis. Students can be trained in lateral reading, cross-source triangulation, and reverse image searches, while librarians and journalists can be invited to conduct workshops on professional verification methods. Fact-checking, when embedded in coursework, becomes more than a technical exercise: it instills a habit of mind that promotes autonomy and resilience.

### 2.5.4 Using real-life ethical scenarios

Simulation-based learning helps students grapple with the ethical dilemmas introduced by GenAI. For example, instructors can simulate grading scenarios in which students submit essays partially written by GenAI, prompting class discussions on integrity, originality, and fairness. Ethical reflection can also be built into rubrics, rewarding transparency and responsible disclosure of AI use. UNESCO's ethical pillars, such as fairness and accountability, can serve as prompts for group discussions and structured reflections. By confronting these dilemmas in realistic scenarios, students learn to navigate the practical tensions between innovation and integrity.

### 2.5.5 Designing transparent AI-augmented assignments

Prohibiting AI tools outright may encourage covert use. A more effective approach is to design assignments that normalize disclosure. Students may be required to note explicitly which parts of their work are AI-assisted, use GenAI only for brainstorming or early drafting, or submit reflective essays comparing their writing with AI-generated alternatives.

These approaches help students see AI as a tool to be critically engaged with, not a substitute for thinking, while cultivating habits of transparency and ethical metacognition.

### 2.5.6 Supporting educators through institutional resources

Teachers cannot implement these strategies without institutional support. Universities should provide professional development on AI pedagogy, prompt design, and AI-assisted feedback. They can share sample syllabi, rubrics, and policies that clarify what constitutes acceptable AI use, while establishing clear institutional guidance on privacy, security, and data protection in AI tools. Such measures reinforce the governance layer of the EAM by ensuring that educators are empowered to lead responsibly and consistently.

### 2.5.7 Addressing the psychological impact of GenAI

GenAI also introduces emotional and psychological challenges. Prolonged exposure to synthetic media can cause epistemic paralysis, where students become skeptical of all information (Fallis, 2021). Others experience AI anxiety, fearing replacement by AI systems or worrying about being misjudged by them (Selwyn, 2023). A further risk is overreliance, where students outsource too much analytical work to AI, leading to diminished autonomy (Holmes et al., 2022). Educators can mitigate these effects by facilitating open discussions about student concerns, by comparing AI- and human-generated work to highlight creativity and judgment, and by encouraging reflective journaling on students' own AI use. These practices help restore trust in human agency while preparing learners to work critically alongside AI.

### 2.5.8 Global case snapshots

Educational responses to GenAI vary globally, offering lessons for adaptation. In Finland, project-based learning integrates AI ethics into civic education (OECD, 2019). In Singapore, AI tutors support STEM education but always under teacher oversight, highlighting a balance between automation and human agency (Touretzky et al., 2019). In Kenya, students produce community radio shows and posters to raise awareness about deepfakes, linking literacy with civic activism (Holmes & Miao, 2023). In Québec, the PIA Trousse pédagogique is used in philosophy and communication courses, sparking debates on bias and misinformation (Bruneault et al., 2024). These cases illustrate that while strategies must be adapted to local contexts, the shared goal remains cultivating resilience, critical inquiry, and ethical awareness among learners.

## 2.6 EMBEDDING ETHICS AND AI LITERACY IN THE CURRICULUM

While teaching strategies address classroom-level practices, long-term impact requires integration into curriculum design and institutional policy. Embedding ethics and AI literacy across programs ensures that GenAI becomes a tool for empowerment rather than a driver of erosion in educational values.

### 2.6.1 Cross disciplinary integration

AI literacy should be embedded across disciplines in the same way that critical thinking or digital literacy has been mainstreamed. Institutions can introduce mandatory AI ethics modules in general education programs to give all students foundational knowledge. In addition, discipline specific-case studies can contextualize risks and applications, for example, examining bias in hiring algorithms in business courses or analyzing how GenAI reshapes historical research in the humanities. Competencies should also be scaffolded progressively, from introductory courses in the first year to advanced research-based applications in senior years. This approach prevents AI literacy from being siloed in technical fields and establishes it as a core competence for all graduates.

### 2.6.2 Revising academic integrity policies

Traditional definitions of plagiarism are insufficient for the GenAI era. Institutions need to update integrity policies to explicitly address AI use by clarifying what constitutes acceptable versus unacceptable practices. Disclosure of GenAI tools should be required in all assignments and research projects, and codes of conduct must include consequences for undisclosed or inappropriate AI use. The UM6P AI Use policy (2024) offers a practical model, requiring students to indicate both the type and version of tools employed, with expectations embedded directly into course syllabi. Such transparency safeguards academic integrity while normalizing responsible AI use.

### 2.6.3 Faculty development and institutional support

Educators play a central role in shaping AI ethics in practice, yet many require training and support. Institutions should, therefore, invest in professional development workshops on AI pedagogy, prompt design, and AI-assisted feedback. Creating communities of practice where faculty can share experiences and strategies strengthens institutional learning. Involving faculty in co-developing policies fosters bottom–up ownership, ensuring that rules are not perceived as externally imposed but rather as part of a shared academic culture.

### 2.6.4 Ensuring equity of access

AI adoption can exacerbate inequalities if not managed carefully. Wealthier students may have access to premium tools, while others are left with free platforms that often monetize user data in ethically problematic ways. Institutions should provide licensed, secure AI tools for all learners to ensure fairness. Additionally, resources must reflect linguistic and cultural diversity to prevent the marginalization of underrepresented groups. Equity is, thus, not only about access to tools but also about ensuring that the tools themselves support inclusive learning.

### 2.6.5 Institutional AI readiness

Embedding AI literacy into curricula requires systemic preparation. Universities should establish AI Ethics and Governance Committees to oversee adoption and ensure alignment with institutional values. Annual impact audits of curricula and assessment practices can help identify emerging risks and unintended consequences. Research protocols should also mandate AI disclosure, requiring scholars to specify the version and function of tools used (Chen & Shu, 2024). These measures align higher education with international publishing standards while reinforcing research integrity and accountability.

### 2.6.6 Partnership and community engagement

AI ethics and literacy are not confined to higher education. Partnerships with schools, non-governmental organizations, media organizations, and policymakers can extend AI literacy across society. Universities can contribute to public awareness campaigns on deepfakes and misinformation, positioning themselves as leaders in shaping ethical digital cultures. In this way, embedding AI ethics does not become a one-time curriculum fix but an institutional and societal commitment to building resilience against misinformation.

## 2.7 CONCLUSION AND FUTURE OUTLOOK

Chapter 2 examined both the promises and the perils of GenAI in education, with a specific focus on misinformation. The analysis demonstrated that misinformation generated or amplified by LLMs and synthetic media presents not only technical risks but also profound ethical, cognitive, and institutional challenges. From hallucinated citations to deepfakes and the blurring of authorship, these risks undermine trust, threaten academic integrity, and erode learner autonomy if left unaddressed.

This chapter yielded several key findings. First, drawing on Keefe (2019), it introduced a typology of misinformation – accidental, ignorant, and intentional – that highlights the need for differentiated responses. Accidental errors demand verification and literacy, ignorance calls for scaffolding and curricular reform, and intentional disinformation requires institutional and sometimes legal action. Second, the review of global frameworks – including those from UNESCO, the European Commission, Stanford University, the PIA toolkit, the OECD, and Canada's FASTER Framework – revealed a strong consensus around principles such as transparency, fairness, human oversight, and accountability. Yet most of these frameworks treat misinformation only indirectly rather than as a primary design target. Third, it proposed an EAM as its unique contribution. The EAM synthesizes these frameworks into a three-layered structure that explicitly links pedagogy, governance, and technical safeguards to the problem of misinformation, thereby bridging the gap between abstract ethical commitments and classroom-level interventions. Finally, the chapter outlined practical strategies for combating misinformation, including embedding AI literacy, training students to detect deepfakes, cultivating fact-checking habits, and designing transparent AI-augmented assignments. When combined with institutional policies and governance, these practices create a multilayered defense against misinformation.

Chapter 2 contributes to the field in three important ways. It reframes information as a central challenge of GenAI in education rather than a peripheral concern, advances a new integrative model (EAM) that aligns ethical principles with actionable responses across multiple levels of practice, and demonstrates, through global case examples, that although strategies vary across contexts, the shared priority remains cultivating resilience, critical thinking, and ethical responsibility among learners.

At the same time, it has certain limitations. As a conceptual synthesis, it does not provide empirical evaluation of the EAM in classroom settings. Future research should pilot these strategies in diverse educational contexts and measure their effectiveness in improving AI literacy and reducing susceptibility to misinformation. More work is also needed to explore how cultural, linguistic, and socioeconomic factors shape students' vulnerabilities to AI-driven misinformation and how equity can be ensured in the deployment of AI tools.

Looking forward, educational institutions must invest in capacity building. Teacher training, curriculum redesign, and governance structures should integrate AI ethics as a core element rather than as an add-on. Equally, partnerships beyond higher education – with schools, civil society organizations, and media outlets – will be essential for building a broader culture of digital resilience. Ultimately, the future of GenAI in education will not be determined by the technology itself but by the values and practices that guide its use. If implemented with transparency, fairness, and human agency at the center, GenAI can enhance learning while strengthening academic integrity. If neglected, however, it risks eroding the very foundations of trust, critical inquiry, and autonomy on which education depends.

## REFERENCES

Abid, A., Farooqi, M., & Zou, J. (2021). Persistent anti-Muslim bias in large language models. arXiv preprint arXiv:2101.05783. https://arxiv.org/abs/2101.05783

Bender, E. M., Gebru, T., McMillan-Major, A., & Shmitchell, S. (2021). On the dangers of stochastic parrots: Can language models be too big? *Proceedings of the 2021 ACM Conference on Fairness, Accountability, and Transparency*, Toronto, Canada, 610–623. https://doi.org/10.1145/3442188.3445922

Bitton, D. B., Hoffmann, C. P., & Godulla, A. (2024). Deepfakes in the context of AI inequalities: Analysing disparities in knowledge and attitudes. *Information, Communication & Society*, 28(1), 1–21.

Bruneault, F., Sabourin Laflamme, A., Boivin, J., Grondin-Robillard, L., & Le Calvez, É. (2024). Former à l'éthique de l'IA en enseignement supérieur: Trousse pédagogique. Pôle interordre de Montréal. https://collimateur.uqam.ca/collimateur/former-a-lethique-de-lia-en-enseignement-superieur/

Chen, C., & Shu, K. (2024). Can LLM-generated misinformation be detected? *Proceedings of the International Conference on Learning Representations (ICLR)*. Vienna, Austria. https://arxiv.org/abs/2309.13788

Chesney, R., & Citron, D. K. (2019). Deepfakes and the new disinformation war. *Foreign Affairs*, 98(1), 147–155.

Doss, C., Mondschein, J., Shu, D., Wolfson, T., Kopecky, D., Fitton-Kane, V.A., Bush, L., Tucker, C. (2023). Deepfakes and scientific knowledge dissemination. *Scientific Reports*, 13(1), 13429.

European Commission. (2022). Ethical guidelines on the use of artificial intelligence (AI) and data in teaching and learning for educators. Publications Office of the European Union. https://op.europa.eu/en/publication-detail/-/publication/d81a0d54-5348-11ed-92ed-01aa75ed71a1/language-en

Fallis, D. (2021). The epistemic threat of deepfakes. *Philosophy & Technology*, 34(4), 623–643.

Floridi, L., & Cowls, J. (2019). A unified framework of five principles for AI in society. *Harvard Data Science Review*, 1(1). https://doi.org/10.1162/99608f92.8cd550d1

Government of Canada. (2025). Guide on the use of generative artificial intelligence in federal institutions. https://www.canada.ca/en/government/system/digital-government/digital-government-innovations/responsible-use-ai/guide-use-generative-ai.html

Holmes, W., Bialik, M., & Fadel, C. (2022). *Artificial Intelligence in Education: Promises and Implications for Teaching and Learning* (2nd ed.). Center for Curriculum Redesign.

Holmes, W., & Miao, F. (2023). Guidance for generative AI in education and research. UNESCO Publishing. https://doi.org/10.54675/EWZM9535

Keefe, T. E. (2019). Not all "fake news" is equal: How should higher education respond to fake news in the post-truth era? *The Liminal: Interdisciplinary Journal of Technology in Education*, 1(1), Article 3. https://digitalcommons.du.edu/theliminal/vol1/iss1/3

Lucy, L., & Bamman, D. (2021). Gender and representation bias in GPT-3 Generated Stories. In *Proceedings of the Third Workshop on Narrative Understanding*, pages 48–55, Association for Computational Linguistics, Virtual. https://aclanthology.org/2021.nuse-1.5

Marcus, G., & Davis, E. (2019). *Rebooting AI: Building Artificial Intelligence We Can Trust*. Vintage.

OBVIA. (2024). Éduquer contre la désinformation amplifiée par l'IA et l'hypertrucage: Une recension d'initiatives de 2018 à 2024. Observatoire international sur les impacts sociétaux de l'IA et du numérique. https://www.obvia.ca/sites/obvia.ca/files/ressources/202410-OBV-Pub-Eduquer_Contre_Desinformation_0.pdf

OECD. (2019). Recommendation of the council on artificial intelligence. OECD Legal Instruments. https://legalinstruments.oecd.org/en/instruments/OECD-LEGAL-0449

Osborne, J., & Pimentel, D. (2023). Science education in an age of misinformation. *Science Education*, 107(3), 553–571. https://doi.org/10.1002/sce.21790

Passi, S., & Vorvoreanu, M. (2022). Overreliance on AI literature review. *Microsoft Research*, *339*, 340. https://www.microsoft.com/en-us/research/publication/overreliance-on-ai-literature-review/

Porayska-Pomsta, K., Holmes, W., & Nemorin, S. (2024). The ethics of AI in education. arXiv preprint arXiv:2406.11842. https://arxiv.org/abs/2406.11842

Selwyn, N. (2023). Should robots replace teachers? AI and the future of education. *Learning, Media and Technology*, 48(1), 1–12.

Stanford Teaching Commons. (2025). Understanding AI literacy. Stanford University. https://teachingcommons.stanford.edu/teaching-guides/artificial-intelligence-teaching-guide/understanding-ai-literacy

Touretzky, D. S., Gardner-McCune, C., Martin, F., & Seehorn, D. (2019). The AI4K12 initiative: Developing guidelines for AI literacy in K–12. *Proceedings of the AAAI Conference on Artificial Intelligence*, 33, 9795–9799.

U.S. Copyright Office. (2023). Copyright registration guidance: Works containing material generated by artificial intelligence. https://www.federalregister.gov/documents/2023/03/16/2023-05321/copyright-registration-guidance-works-containing-material-generated-by-artificial-intelligence#h-9

UNESCO. (2021). Recommendation on the ethics of artificial intelligence. United Nations Educational, Scientific and Cultural Organization. https://unesdoc.unesco.org/ark:/48223/pf0000381137

University Mohammed VI Polytechnic. (2024). AI use policy for academic and research settings. https://policy.um6p.ma/knowledgebase/ai-use-policy/

Yan, L., Sha, L., Jin, Y., & Echeverria, V. (2024). Generative AI in higher education: A global perspective of institutional adoption policies and guidelines. *Computers and Education: Artificial Intelligence*, 8, 100348. https://www.sciencedirect.com/science/article/pii/S2666920X24001516

Part II

# Large language models (LLMs) for misinformation detection

Chapter 3

# Zero trust architectures and cybersecurity frameworks for mitigating AI-enabled misinformation campaigns

*Shahed Al-Tamimi and Qasem Abu Al-Haija*

## 3.1 INTRODUCTION

With the rapid development of technology and artificial intelligence (AI), data privacy and information are vulnerable since no system, software, or network is 100% secure. Thus, there must be trust between the system and users since if there is no trust, the users may use the technology and interact with the intelligent systems within various digital environments. Based on trust, this section introduces the Zero Trust Architecture (ZTA) and briefly overviews the misinformation crisis.

### 3.1.1 Overview of the misinformation crisis in the age of generative AI

To start with defining misinformation, it refers to false or misleading data that can be shared either on purpose or by accident. However, this widespread prevalence of misinformation caused various obstacles that are being amplified by generative AI, which is capable of generating deepfakes, altered texts, and manipulated media. Generative AI is a captivating aspect of AI that grows from a lot of data in order to generate novel content comprising images, music, video, or text. It has made very rapid development with the introduction of Large Language Models (LLMs) and improved Generative Adversarial Networks (GANs). New technologies have benign applications in innovation, automation, and problem-solving; they also enable producing realistic but deceptive content on a large scale. To deal with this crisis, some requirements need to be set: reliable detection mechanisms, ZTA security, transparency, content authentication, and AI-based systems to guarantee that digital communication contexts are genuine, trustworthy, and resilient (Ejeofobiri et al., 2022). Indeed, the emerging threat landscape is defined by integrating AI into disinformation operations, which showed the manner in which doctored narratives and AI-generated propaganda undermined democratic procedures and eroded social cohesion (Olanipekun, 2025). Also, conventional ways of defending against these dangers are not enough since they are too fast and smart. Consequently,

DOI: 10.1201/9781003626558-5 

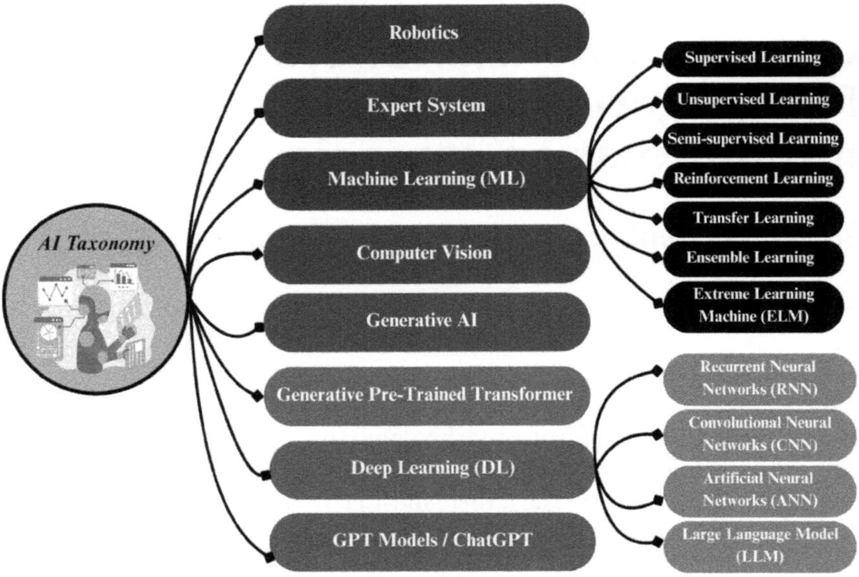


*Figure 3.1* AI taxonomy.

the deployment of adaptive cybersecurity frameworks grounded on Zero Trust (ZT) principles and AI-driven threat detection was essential for the successful identification, mitigation, and containment of AI-enabled disinformation campaigns. Figure 3.1 shows how AI technology is divided into two groups: innovation and manipulation.

### 3.1.2 Limitations of perimeter-based security in combating misinformation dissemination

Perimeter-based security, traditionally viewed as the cornerstone of organizational defense, is becoming less effective in addressing the spread of misinformation in the contemporary digital environment. However, AI models are predicated on the outdated assumption that attacks mostly originate from external sources, while considering all components inside the network perimeter as fundamentally reliable (Paul et al., 2024). Misinformation often capitalizes on hacked internal accounts, trusted individuals, and social engineering, rendering static perimeters inadequate. The transformation across cloud-first infrastructures, distributed networks, and hybrid ecosystems indicates that data and misinformation campaigns can spread without crossing traditional network perimeters. Additionally, the attackers can exploit the agility of AI-driven tools to develop dynamic, adaptive misinformation environment that easily bypasses signature-based detection and static controls, as well as the result of perimeter-based defenses' lack of continuous verification, contextual awareness, and adaptability required to combat the evolving scale and sophistication of AI-enabled misinformation (Ofili et al., 2025; Tiwari et al., 2022).

**The key limitations include the following:**

- Static defense constraints failed to stop criminals from spreading false information.
- Performance is limited to cloud and hybrid settings.
- Capacity storage is limited to adjust to AI-driven and changing misinformation approaches.
- There is an absence of detailed ongoing algorithms.
- Adversaries can readily exploit overreliance on the distinctions between external and internal trust.

### 3.1.3 Importance of cybersecurity frameworks

The importance of ZT in content authenticity, platform security, botting suppression, and next-generation AI-powered disinformation campaigns underscores the need for stringent cybersecurity architectures capable of handling content legality, platform integrity, and automated threats like malicious bots. Adaptive frameworks based on ZTA represent a central alternative to collapsed perimeter-based models. However, Ejeofobiri et al. (2022) emphasize that adaptive cybersecurity architectures based on ZT models and AI-driven threat detection provide continuous authentication and dynamic threat response to ensure that even internally generated misinformation can be checked before it is disseminated. Constant testing increases the authenticity of the content. However, the ZT platform enhances security with cloud-based and AI-powered security measures by reducing implicit trust and enforcing identity verification (Celeste & Michael, 2021). Thus, digital platforms are prone to large-scale disinformation and bot-driven amplification efforts to benefit from these strategies. AI and ZTA synergize within next-generation frameworks, and disinformation actors increasingly use advanced AI technologies (Paul et al., 2024). AI integration is key to providing adaptive controls that identify and neutralize false stories in real-time across the ZTA framework. Further, Ofili et al. (2025) demonstrate how ZT, AI-driven threat intelligence, and compliance approaches can boost government cloud infrastructures, where the discussed frameworks that prevent bots and disinformation tools from compromising sites always verify and contextualize access requests. Further integrating AI into ZTA (Tiwari et al., 2022) improves adaptive security, enabling businesses to identify developing misleading techniques. Even though generative AI–driven attacks can modify quickly for conventional systems, flexibility is essential. Lastly, Al-Tamimi et al. (2024) emphasize the relevance of ZTA in safeguarding Internet of Things (IoT) networks, which are being used to spread disinformation via hacked devices. IoT and edge systems using ZT principles provide complete protection for digital ecosystems.

### 3.1.4 Chapter objectives and scope

This chapter aims to identify the most recent cybersecurity frameworks for ZTA to combat AI-driven misinformation campaigns. Recent research indicates that AI and ZTA may converge to facilitate adaptive, continuous verification and dynamic threat response within complex digital ecosystems (Tiwari et al., 2022). Furthermore, the scope extends to securing IoT networks, illustrating how ZT principles may defend dispersed and linked systems against AI-driven disinformation assaults (Al-Tamimi et al., 2024).

### 3.1.5 Chapter organization

This chapter provides an overview and discusses the role of mitigating AI, which enabled misinformation campaigns for ZTA and cybersecurity frameworks. Section 3.1 provides the introduction. The subsequent sections of this chapter are structured as follows. Section 3.2 illustrates the concept of AI-enabled misinformation campaigns. The characteristics of AI-enabled misinformation and attack surface via digital platforms are also discussed. Section 3.3 provides ZTA principles and foundations. Section 3.4 identifies ZT to misinformation mitigation approaches and the possible ways to employ them. Section 3.5 describes cybersecurity frameworks and standards. The integrated defense strategies, tools, and technologies are identified in Section 3.6, and the case studies and applications are explored in Section 3.7. The challenges and future directions are provided in Section 3.8. Finally, the conclusion of the chapter is provided in Section 3.9. In short, this chapter provides substantial and well-documented information.

## 3.2 UNDERSTANDING AI-ENABLED MISINFORMATION CAMPAIGNS

The AI-enabled misinformation campaign has proven to be an important threat to the digital ecosystem and utilizes generative AI skills, automatic material production, and advanced distribution methods. Unlike traditional resolutions, these campaigns benefit from machine learning (ML), LLM, and synthetic media to create solid materials, propagate them quickly, and influence the target audience. Designing an effective cybersecurity structure is necessary to understand its characteristics and attack surfaces, especially based on ZT principles (Gudala et al., 2021).

### 3.2.1 Characteristics of AI-enabled misinformation

A. **Automated content creation (e.g., deepfakes and GPT-generated text):** Generic AI technologies, which consist of deepfakes that generate fake videos utilizing GPT-based synthesis to generate various texts and images, have enabled automatic materials on a unique scale and with unprecedented realism. These gadgets can create illusions, manipulate such images, or produce synthetic sounds to mislead viewers. Besides this, the automating content production generated by generic AI technologies increases the volume, variety, and sophistication of misinformation, posing obstacles to traditional moderation and verification systems. However, the integration of AI through the zero confidence framework facilitates real-time detection and adaptive mitigation, making it possible to identify suspicious material. (Gudala et al., 2021; Kolawole, 2025; Muriithi et al., 2025).

B. **Rapid propagation through botnets and fake accounts:** AI skilled misinformation campaigns utilize botnets and virtual narratives to spread content rapidly in social media and online networks. Automated amplification tactics maximize exposure to network effects along with algorithmic bias in the recommended engines. Due to this, various initiatives are continuously introduced to prioritize dynamic verification as well as authentication methods within ZTA, bypassing the conventional perimeter security (Ajish, 2024).

C. **Targeted psychological manipulation and narrative shaping:** In addition, the dissemination of information that encompasses AI-generated disinformation is typically designed to target particular demographic groups, shape circumstances, and manipulate views. Founded on behavioral insights and segmentation of AI that worked into target audiences, such campaigns propagate personalized messages that invoke cognitive bias and emotional stimulus, where advanced cybersecurity systems that use AI and ZTA can monitor anomalous commitment patterns, detect coordinated influence campaigns, and counter the psychological effects of targeted disinformation (Liu et al., 2024; Salem et al., 2024).

### 3.2.2 Attack surface in digital platforms

A. **Social media manipulation:** Social media platforms exhibit significant susceptibility to AI-generated misinformation due to user-generated content and dependence on algorithmically curated feeds. Contemporary actors use popular themes, hashtags, and engagement metrics to enhance misleading and fabricated narratives. Consequently, ZT frameworks use stringent authentication, continuous monitoring, and deviations to prevent the illegal or harmful amplification of content (Ud Din et al., 2024).

B. **Abuse of content recommendation systems:** Latest recommendation algorithms which are discussed in Gudala et al. (2021); Kolawole (2025) to get end-users to misinformation might unintentionally help AI-generated misleading information to propagate. These arrangements can lead to misinformation that seems more credible as it focuses on individuals with similar interests or by creating echo chambers. Adding AI-powered hazard detection to ZT frameworks can analyze how material spreads in real-time, spot suspect amplification, and stop misinformation from spreading too far (Gudala et al., 2021; Kolawole, 2025).

C. **Weaponization of LLMs & synthetic accounts:** LLM and fraudulent payments have become tools for planned efforts to spread misleading data, social manipulation, and distribute misinformation. These AI-powered businesses can behave similar to humans, respond straight immediately, and adjust their messages based on what individuals in their target market say. This renders normal security ineffective. ZTA, together with AI analytics, lets you check the behavior of people and tools all the time, which prevents bad LLM-powered debts from hurting the integrity of the platforms (Ajish, 2024; Liu et al., 2024; Salem et al., 2024; Ud Din et al., 2024).

D. **Deceptive content (Rivera et al., 2024):** Finally, the campaign enters a reinforcement stage, where repeated exposure, echo chambers, and algorithmic biases entrench the false narratives, making detection and correction increasingly difficult. Table 3.1 also identifies a summary of the lifecycle.

The life cycle of the AI-enabled misinformation campaign usually begins with the generation, where advanced generative models form misleading materials such as deepfake videos, synthetic sound, or fabricated texts, designed to manipulate perception (Altamimi & Salameh, 2024). A sample life cycle is given in Figure 3.2. This is followed by packaging and adaptation, where malicious actors refine the material to target the specific target groups through linguistic seams, cultural framing, or emotional triggers.

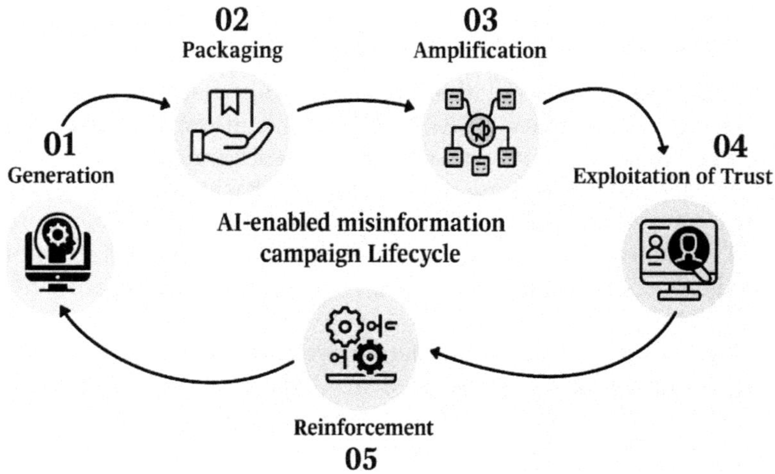


*Figure 3.2* Misinformation campaign AI-enabled lifecycle.

The third phase is reinforcement, where social media is strategically spread through algorithms, bottom networks, or compromised platforms to maximize access and engagement (Muriithi et al., 2025). When it is widely distributed, misinformation utilizes the digital identity and weaknesses of the trust mechanism, which benefits from bad stories or identity theft to legalize misleading materials (Rivera et al., 2024).

*Table 3.1* Summary of AI-enabled misinformation campaign lifecycle

| *Lifecycle step* | | *Description* |
|---|---|---|
| **Step 1** | **Generation step** | The lifecycle begins with generating misinformation content employing advanced AI technologies and models such as GAN and LLMs. To this point, criminals may create deepfake videos, synthetic sounds, or fabricated articles imitating authentic sources. This generation process takes advantage of massive datasets to ensure linguistic fluency, realistic images, and confident stories (Altamimi & Salameh, 2024). |
| **Step 2** | **Packaging and adaptation** | When generated, misinformation is packed for highest possible power. This involved incorporating messages with emotional triggers (fear, anger, and urgency) and sewing the material for specific linguistic, cultural, or political contexts. Attackers also optimize materials for digital platforms, for instance, shortening the video for Ticket Chef, adding hashtags to Twitter/X platform, or forming visuals for Instagram. This phase ensures that false material is more reliable. |
| **Step 3** | **Amplification** | The misinformation package is then reinforced using strategic bot networks, fake accounts, sorcerers, and recommendation algorithms. Automatic post plan paid reproduction or kidnapped trend hashtags are used to push the wrong information in public feed, where it gets viral speed (Muriithi et al., 2025). Repeatedly using the "illusory truth effect" of the explosion creates a false sense of credibility in the exposure volume. |
| **Step 4** | **Exploitation of trust** | After reinforcement, misinformation seems reliable for a digital trust mechanism. This may include spreading through staff or hacked accounts, applying verified profiles, or utilizing weaknesses in digital identity systems. Entering resolution within reliable communication channels sidelines attackers' skepticism and increases the adoption of false stories (Rivera et al., 2024). |
| **Step 5** | **Reinforcement** | In the final phase, incorrect information becomes self-confidence. Algorithm eco chambers, filter bubbles, and social verification loops constantly strengthen false stories. The society is constructed around the principles of conspiracy or conceptual prejudice, and repeated and misinformed information is reopened and continued in opinion. Currently, the mechanisms of improvement face as frequent exposure has normalized lies and the efforts to debate are often rejected as biased or biased by sensors. |

*Table 3.2* Comparison of traditional vs. AI-enabled misinformation attributes

| *References* | *Attribute* | *Lifecycle phase* | *Traditional misinformation* | *AI-enabled misinformation* |
|---|---|---|---|---|
| Gudala et al. (2021), Kolawole (2025) | Content creation | Generation | • Manually written or produced by humans, with limited volume and consistency | Automatic generations using LLMs, deepfakes, GANs, high volume, realistic, and diverse. |
| Kolawole (2025), Aleisa (2025) | Adaptability | Packaging | • Static narratives<br>• Minor<br>• Modifications over time. | Dynamic and adaptive systems evolve in response to employ detection and user feedback. |
| Ajish (2024), Liu et al. (2024) | Propagation speed | Amplification | • Relatively slow, relies on manual sharing or broadcasting. | Rapid, amplified by botnets, fake accounts, and automated posting. |
| Salem et al. (2024), Joshi (2024) | Targeting | Exploitation of trust | • Broad, non-personalized.<br>• General audiences. | Highly targeted, employing AI-driven audience segmentation and behavioral analysis. |
| Gudala et al. (2021), Muriithi et al. (2025) | Content realism | Generation and packaging | • Lower fidelity and easy detection of deepfakes and fake content. | High realism, deepfake media, synthetic text, audio, and images. |
| Ud Din et al. (2024), Rivera et al. (2024) | Attack surface | Amplification and exploitation | • Limited to conventional channels (print, email, and broadcast). | Multi-platform:<br>• Social media<br>• Messaging apps<br>• IoT devices<br>• Recommendation systems<br>• Metaverse |
| Ajish (2024), Joshi (2024) | Detection difficulty | Reinforcement | • Easier to identify via manual fact-checking or traditional filters. | An obstacle to detecting and required for AI-driven anomaly detection and ZT verification. |

The AI-competent incorrect information campaign represents a structured and more sophisticated development of traditional misinformation. While traditional misinformation depends on manual fabrication, rumors, or biased reporting, AI-operated methods such as GAN and LLMS are recreated in the media. The life cycle usually begins with the generation, followed by packaging and framing, reinforcement through robots or social platforms, utilization of faith, and finally reinforcement through frequent exposure (Rivera et al., 2024). Unlike traditional misinformation, which extends slowly and finds, AI-competent misinformation is quickly adapted; the algorithm utilizes reinforcement and reduces the confidence in digital ecosystems. To combat this, ZTA, blockchain-competent verification, and AI-Run Detection Framework have been proposed to secure identity, secure privacy, and discover deepfake manipulation (Aleisa, 2025; Joshi, 2024; Chen et al., 2021). Table 3.2 illustrates the comparison of traditional and AI-enabled misinformation attributes.

## 3.3 ZERO TRUST ARCHITECTURE (ZTA): PRINCIPLES AND FOUNDATIONS

### 3.3.1 What is zero trust?

ZTA has proven to be a paradigm change in information protection and moved away from the traditional "trust but verify" approach to one based on the principle of "never trust, always verify" (Al-Tamimi et al., 2024). Unlike circumference-based security models, which believe that the threats are mainly generated from outside the network, ZTA accepts that the interior can ignore the threats, compromised accounts, and advanced persistent attacks. This philosophy reflects the scenario of changing danger, where attackers can benefit from the weaknesses only since the first reach is achieved. The ZT model states that each request for access to data, app, or services is continuously certified, authorized, and encrypted, even if the request is generated from or out of the organization's network. This ensures no user, device, or system component receives any underlying trust. The ZTA mitigated the risk of lateral movement and privilege attacks through the implementation of minimal-rich accession and continuous verification (Al-Tamimi et al., 2024). Where the increasing complexity of IT ecosystems, marked by cloud computing, proliferation of mobile devices, and IoT integration, has rendered ZT essential for organizations to safeguard sensitive data and critical infrastructure security. Recent research has emphasized the growing role of AI in enhancing ZTA, facilitating rapid threat identification, adaptive policy enforcement, and the detection of real-time deviations (Rana, 2025). It reflects both the advancement of ZT in a prominent safety theory and its dynamic structure. ZTA redefines network security by removing assumptions of implicit trust and implementing continuous verification, micro-segmentation, and stringent access controls.

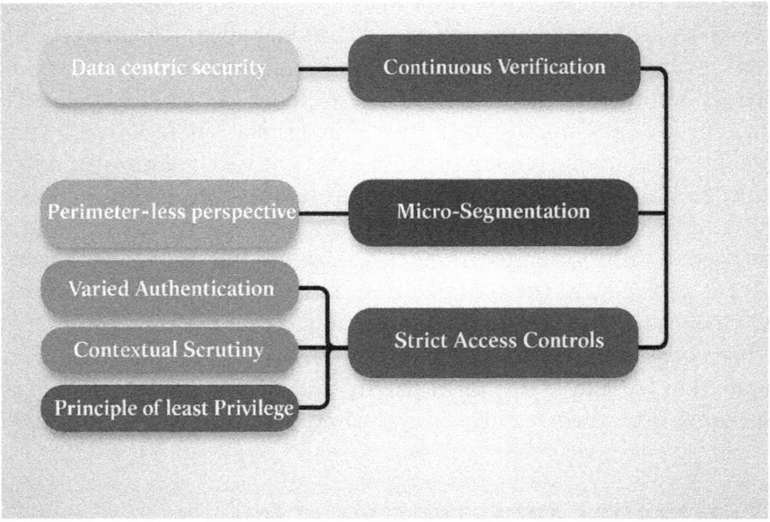


*Figure 3.3* ZTA conceptual map. (From Khan, 2023.)

This highlights the necessity of permanent authentication for users, equipment, and applications, while mitigating side effects caused by network partitions. Complementary principles, including data-centered protection and a circumference wall perspective, ensure that the data remain preserved regardless of the location, and the dangers are examined both in the interior and the exterior. Overall, these foundations install a flexible, adaptive, and dynamic safety structure for the complications of the modern digital environment (Khan, 2023); the ZTA conceptual map is provided in Figure 3.3.

### 3.3.2 ZTA model and components based on the NIST

The most acclaimed and structured representation model of ZTA is proposed by the National Institute of Standards and Technology (NIST), which defines a set of leading principles and main components for effective ZT (Al-Tamimi et al., 2024). According to the NIST, ZT contains various important elements, each working together to ensure a unified and adaptive security currency. The NIST reference architecture is illustrated in Figure 3.4.

- **Policy decision point (PDP):** The central component responsible for making access decisions. It evaluates requests against security policies and risk indicators to determine whether access should be granted.
- **Policy enforcement point (PEP):** Implements access decisions made by the PDP, ensuring that traffic is allowed, denied, or restricted according to policies.

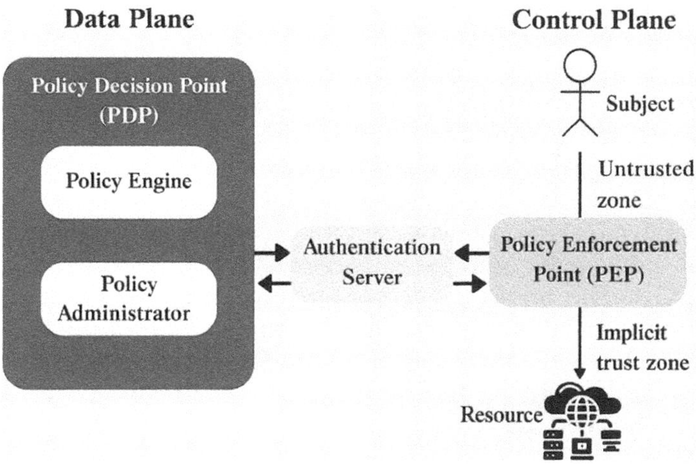


*Figure 3.4* Zero trust reference NIST architecture employed to digital content platforms.

- **Continuous diagnostics and mitigation** (CDM)**:** Provides real-time monitoring of user activities, device health, and network conditions to ensure compliance with ZTA principles.
- **Dynamic policy engine:** Adjusts security policies dynamically in response to changing conditions, such as detected anomalies or user context.
- **Data access governance:** Ensures that access is limited to authorized entities and monitored to detect any attempts at unauthorized data exfiltration.

The NIST reference architecture for the ZTA model emphasizes that the decisions of the threat are not stable. Rather, they should develop with reference to the risk score and observed behavior. On the contrary, older models rely on stable credentials or network rooms for access control. With the increase in AI and automation, these components are rapidly increasing to provide future and adaptable abilities, which reduces the difference between static security structures and modern dynamic threats (Brief, 2021). In addition, AI-driven equipment may increase (Hattali, 2024) before identifying PDP and CDM, correlating equipment deviations, and identifying potential internal formulas. For instance, AI-operated User Behavior Analysis (UBA) can detect deviations from baseline activity, enabling real-time policy adjustment. In this way, the NIST model develops in an AI-driven architecture that uses zero confidence principles and estimates possible attacks.

### 3.3.3 Benefits of ZTA in information security and integrity

ZTA implementation supplies corporations with an extensive range of capabilities that can assist them in coping with modern concerns regarding cybersecurity. The following subsections highlight key advantages supported by recent research.

A. **Protection against lateral movement:** Conventional security approaches that rely on circumference are often vulnerable. When the attackers have early access, they can transfer the network to increase their privileges and exfiltrate data. ZTA eliminates this weakness of limiting access to all steps that require continuous verification and division (Al-Tamimi et al., 2024). Micro-segmentation constitutes one of the most vital components of ZTA. This breaks the network into fragments so that attackers are unable to find their way around the resources on their own. ZTA significantly reduces the attack surface by ensuring that the user and device have only the necessary permissions they need. The AI-enhanced ZTA framework ensures this protection even further by utilizing the ML framework to detect patterns of lateral movement. This can prevent criminal activity before it starts (Rana, 2025). The following proactive approach is an alternative for a defensive approach aimed at ensuring that attackers cannot take benefit of the network's trust all at once.
B. **Real-time access monitoring and enforcement:** ZTA offers attractive benefits that might help ensure and verify honest devices as well as end-users to follow access control rule permissions and settings, where every access request is recorded on the system and revised. However, various devices are utilized depending on various aspects such as attitude, user behavior, location, and time of access (Hatali, 2024). Thus, rendering access choices are based on more than just identification; they are also based on modifying risk factors. For example, if a device shows signs of agreement, the system can immediately interrupt access, even if the user has valid identification. Similarly, if a different login effort is detected from an abnormal geolocation, access requests for further verification can be flagged. However, AI technologies significantly increase this aspect of ZTA by enabling adaptive reactions, where ML algorithms can identify novel cyber risks and alter guidelines without human permission. Thus, real-time adaptability ensures that the organization remains flexible against zero-day attacks and rapidly developed attack vectors.
C. **Role in verifying *users*, devices, and digital agents:** regarding the core of ZT that ensures only certified and authorized institutions can interact with users, equipment, or sensitive systems (Al-Tamimi et al., 2024). The core principle is provided in Figure 3.5. The verification procedures include Multifactor Authentication (MFA), continuous risk assessment, and verification of unit compliance with security policy. Regarding the increasing spread of IoT units and automated digital means, this role of verification has become more important than ever. Various IoT points lack underlying safety facilities, making them a possible entrance point for attackers. ZTA addresses the need for continuous authentication and health checks before allowing a network interaction. In addition,

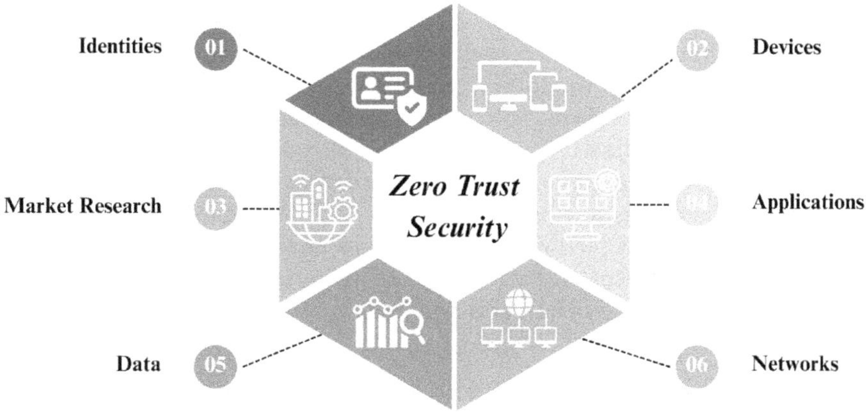


*Figure 3.5* Core principle of secure ZTA.

the AI-operated ZTA improves the confirmation process by assessing relevant information and performing behavioral analysis to distinguish between valid and malicious activity (Mensah 2024). For instance, an AI-driven structural device can detect micro deviations in the communication pattern that avoid traditional rule-based identity. This not only ensures that authorized users and equipment have access to but also that potentially compromised digital agents are discovered and that they are immediately cut. ZTA examines the most significant regulations that may boost the cybersecurity domain as well as replace ZT that exists with continuous verification, adaptable policy enforcement, and relevant decision-making. Even though the rises of the NIST framework identify structured foundation for ZTA, AI integration evolves the effectiveness of modern cyber threats. However, there are significant various positive aspects which include limiting side movement, monitoring probability, and strengthening users, equipment, and digital agents. In the modern age of dynamic cyber environments, where cyber risk is adaptive, varied, and swiftly enables ZTA, it provides a robust and secure paradigm that is well prepared for the future. Its adaptation with minimum privileges, partitions, and continuous monitoring ensures that its alignment with companies and the IoT environment ensures integrity, privacy, and flexibility.

## 3.4 APPLYING ZERO TRUST ARCHITECTURE (ZTA) MISINFORMATION MITIGATION

Rapid growth in misinformation, especially when AI-competent materials work in production and reinforcement, presents a new challenge for cybersecurity and information integrity. Traditional rescue, such as circumference-based control, material moderation, or insulated fact-oriented mechanisms, often proves inadequate to reduce highly adaptable and automated misinformation campaigns. ZTA, with the principle of "ever trust, always confirm," provides a promising observation to address these dangers. Organizations can create flexible systems for user and device access to ZTA concepts by transporting content, verification, and dissemination processes. This section explains how ZTA principles can be operated for misinformation depletion in three main domains: material verification and source confidence, API gateway and material delivery control, and AI-operated traffic and behavioral analysis.

### 3.4.1 Content verification and source trust

The first step of misinformation under a ZT pattern is to address the reliability of the material at the source. In ZTA, access control applies not only to users and equipment but also to information manufacturers and their digital outputs (He et al., 2022). Material manufacturers should be subject to controlling individual journalists, social media accounts, or automatic bot identities' individual access, which verifies their validity before providing publishing rights. By implementing multiproof authentication and verification of cryptographic identity, organizations can ensure that only certified and authorized institutes produce and distribute information (Syed et al., 2022). Another important ZTA mechanism is the material signature and watermark. Just as the ZTA network trust depends on the digital certificate, it similarly requires limiting incorrect information in the cryptographic verification of material authenticity. Intrinsic ZTA framework can detect materials on a conformed signature, digital watermark, or blockchain-based Provence Tag target groups and platforms on a legitimate origin (Ahmadi, 2024). This prevents malicious actors from injecting the content produced in an anonymous ecosystem. In addition, ZT principles can make continuous confirmation of sources compulsory. Unlike static bleaching of reliable media, the ZTA behavior pattern and deviation detection (Nahar et al., 2024) validate the publisher as a dynamic reliability based on Nahar et al. (2024). For instance, if a historically reliable account suddenly begins distributing inconsistent sensational or AI-related materials with its previous behavior, the ZTA system can flag or argue that material. This adaptive verification account strengthens rescue against kidnapping and internal formula hazards. By integrating identity-aware access controls with cryptographic attestation and dynamic trust validation, ZT ensures that the integrity of content is preserved from the moment of creation, minimizing the ability of malicious actors to masquerade as legitimate sources of information.

### 3.4.2 API gateways and content delivery controls

When the content is generated, the misinformation campaigns trust too much in automated distribution systems, including Botnets, fake social media accounts, and programmatically generated posts. ZT principles to disrupt this distribution chain can be extended to the API gateway and the material distribution duct. Under ZTA, all API interactions are subject to minimized access control and continuous monitoring (Khan, 2023). Instead of allowing unlimited posts or material enhancement, platforms can use granular permissions to establish identity. For instance, being forbidden may be forbidden in newly created accounts or rejected institutions, but it establishes confidence through the faith's stability. Speed limit and behavioral monitoring represent another ZTA-induced security. Traditional rescue often uses stable posts, but incorrect information campaigns may be suitable for bypassing them. Conversely, the ZTA-operated systems continuously monitor the account's behavior and use adaptive gas. Accounts showing features such as bot-like mail, synchronized activity, or known resolution groups can be slow, exposed, or suspended (Edo et al., 2022). In addition, ZT concepts enable reference-incredible material delivery control. The way the ZTA unit prohibits network current depends on the level of attitude and user risk; incorrect information can filter material proliferation based on the reliability of both the ZTA system source and the receiving environment. For example, sensitive incorrect information on choices can be prevented from large-scale reinforcement if the source has not gone through full confidence. The API gateway and delivery control gradually work under zero confidence principles, acting as choke points against automated distribution. This ensures that even if malicious content is generated, its ability to spread widely is constrained.

### 3.4.3 AI-driven traffic and behavior analysis

While the material verification and distribution control the incorrect information on the source and transfer layers, continuous monitoring of traffic and user behavior represents the final layer of the ZT applications. In traditional ZTA implementation, traffic inspection and micro-blocks prevent side weights in a network. For incorrect information, this behavior is translated into biometrics and AI-operated analysis that distinguishes between authentic human activity and automated resolution campaigns (He et al., 2022). An AI-enabled ZT system can analyze account activity patterns, including login frequency, the unit's fingerprint, language use, and temporary insertion to detect. For example, suppose an account shows sudden geographical changes in the login site or on a specific day around automation, in that case, the ZTA verification teams can authenticate again or use sanctions (Syed et al., 2022), which may interfere with the claim. Behavioral biometry also expands detections at interaction levels such as key-pressure dynamics, mouse movements, or touch screen movements, which can help human users differ from robots (Ahmadi, 2024). While associated alongside detecting network traffic, this approach developed a multilevel verification system which minimizes the dependence on unsecured static identity rules for the theft of AI-enabled misconception systems (Nahar et al., 2024). ZTA AI-enhanced frames may benefit from threat intelligence to refine detection continuously. As the misinformation strategy developed, the system integrates novel behavioral indicators, such as ZTA updates access policy based on persistent risk reflections (Khan, 2023). This ensures that misinformation defenses remain dynamic rather than static. Additionally, ZT emphasizes real-time enforcement rather than post-incident remediation. Continuously analyzing traffic patterns and user behavior, ZTA systems might block or quarantine misinformation before it reaches large audiences (Edo et al., 2022). This practical approach mirrors the broader shift in cybersecurity from reactive defense to pre-emptive resilience. Integration of ZT controls into platform-level misinformation defenses is shown in Figure 3.6. Table 3.3 represents the mapping of ZTA principles to misinformation mitigation strategies. Applying ZT principles for misinformation mitigation leads to constant verification and control of one of the problems, thus leading to improvement.

Through material verification and source confidence, ZTA prevents unauthorized or fabricated material from entering the ecosystem. Strengthening the API gateway and material delivery control interferes with the cured reinforcement and ensures that only reliable institutions can distribute one scale. Finally, through the AI-operated traffic and behavioral analysis, the

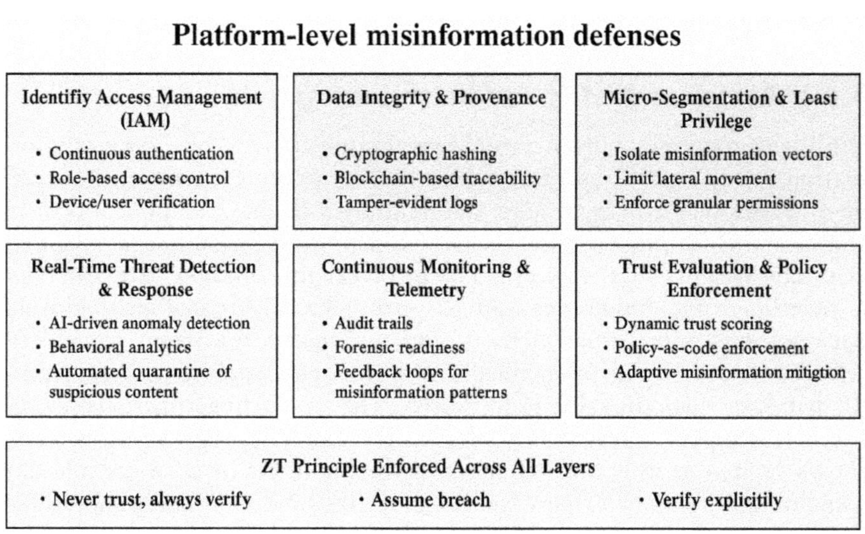


*Figure 3.6* Integrating ZT controls into platform-level misinformation defenses.

*Table 3.3* Mapping ZTA principles to misinformation mitigation strategies

| *References* | *ZTA principle* | *Misinformation mitigation strategy* |
|---|---|---|
| • Ejeofobiri et al. (2022)<br>• Gudala et al. (2021)<br>• Altamimi and Salameh (2024) | Continuous and daily verification | Real-time validation of user identification and content authenticity.<br>Detect deepfake accounts and bot-driven disinformation. |
| • Tiwari et al. (2022)<br>• Liu et al. (2024)<br>• Muriithi et al. (2025) | Micro-segmentation | Isolating malicious botnets or coordinated misinformation clusters to prevent lateral spread across platforms |
| • Paul et al. (2024)<br>• Rivera et al. (2024) | Strict access controls (least privilege) | Limiting content creation or propagation privileges.<br>Ensuring contextual scrutiny of media uploads.<br>Reducing insider-driven misinformation. |
| • Aleisa (2025)<br>• Ofili et al. (2025)<br>• Romanishyn et al. (2025) | Data-centric security | Embedding encryption, watermarking, and provenance checks into media assets to secure integrity and trace manipulation. |
| • Syed et al. (2022)<br>• Chen et al. (2023)<br>• Kolawole (2025) | Perimeter-less perspective | Universal verification of internal and external sources; addressing misinformation from compromised internal accounts or external fake identities. |
| • Olanipekun (2025)<br>• Salem et al. (2024)<br>• Muthusamy (2025)<br>• Ajish (2024) | AI-enhanced threat detection (cross-cutting) | Leveraging AI/ML models to detect coordinated inauthentic behavior, narrative laundering, and AI-generated propaganda. |

ZTA constantly monitored the information environment and identified and lowered real-time deviations. These measures suggest that zero confidence, originally designed for Enterprise Cyber Security, can be a strong structure to manage the developed threat of AI-competent incorrect information. The adaptive, identified, and constant verification theory is essential for protecting information integrity in a digital scenario from automation and manipulation.

## 3.5 CYBERSECURITY FRAMEWORKS AND STANDARDS

To successfully construct misinformation mitigation operations, cybersecurity frameworks may be used, which deal with unique mitigation strategies to tackle integrity issues. Several frameworks, including the NIST framework, offer a systematic method for identification, monitoring, and defense against malicious activities. Notable examples include the NIST Cybersecurity Framework (CSF), MITRE ATT&CK, and ISO/IEC 27001. The recent literature highlights the importance of broadening ZTA concepts to enhance these frameworks (Fernandez & Brazhuk, 2024).

### 3.5.1 NIST cybersecurity framework (CSF)

NIST CSF offers a comprehensive guideline consisting of five core functions: Identify, Protect, Detect, Respond, and Recover, where each function could be adapted for misinformation mitigation for incidents (Chen et al., 2023). Figure 3.7 identifies NIST CSF to provide a clear overview of the core five functions.

- **Identify:** To identify misinformation, within organizations, institutions must first install a baseline with reliable sources, platforms, and communication channels, where the identification phase involves sorting features as well as assessing information system weaknesses and vulnerabilities and evaluating data pipeline reliability. As an illustration, social media platforms that are owned by institutions may recognize the organized network of AI-generated content narratives intended to enhance disruptive tales.
- **Protect:** After identifying vulnerabilities, threats, and weaknesses, the protection involves implementing access control, encryption, antivirus, and a digital watermark, as well as another protection approach, algorithm to ensure data integrity based on the requirement.
- **Detect:** During CSF, the detection demands data flow, traffic patterns, and rigorous oversight of discrepancies in account interactions and databases, which involve utilizing ML models for the real-time detection of malicious activities as well as bot-driven activities and disruptive patterns (Sharma, 2022).

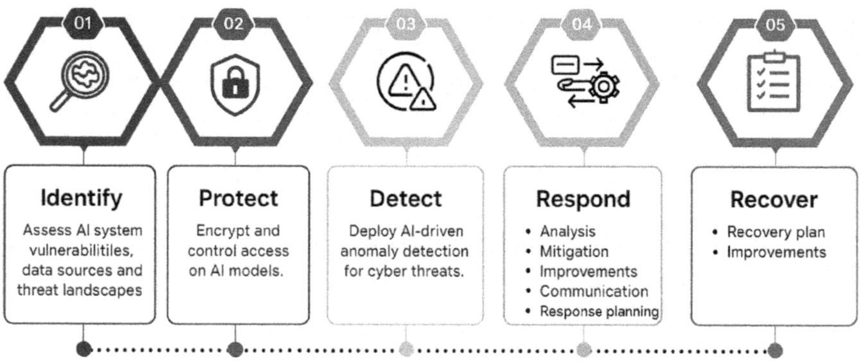


*Figure 3.7* Overview of the NIST cybersecurity framework (CSF) core functions for AI risk management.

- **Respond:** While misinformation campaigns are detected, it must be a built-in reaction function that ensures timely intervention through taking down misinformation and notifying stakeholders about warnings and mitigation steps to protect stakeholders' sensitive information as well as financial data to prevent any loss. Thus, the respond function should consist of a public communication strategy to maintain trust.
- **Recovery:** The recovery function emphasizes the restoration of overall operations, data, and dependability from backed-up servers or drives. Subsequently, detecting the misinformation and source, businesses may disseminate confirmed further enhancements, conduct audits, and bolster adaptive strategies for future occurrences.

Latest applications of CSF allow the organizations to integrate misinformation detection and response mechanisms within their cybersecurity posture, bridging traditional IT security systems with content authenticity. Furthermore, employing CSF to detect misinformation incidents acts as a content security assurance lifecycle when tailored specifically for misinformation detection. Ultimately, the identify function assesses sources of fake narratives and vulnerabilities, detection function observes fake narratives, while the respond function either eliminates or indicates vulnerabilities. Further, the recover function helps rebuild confidence and provides users with material that is backed up by proof (Din et al., 2024). This cycle ensures that the system not only combats misinformation but also continually improves its self-protection mechanisms.

### 3.5.2 MITRE ATT&CK for disinformation operations

The MITRE ATT&CK framework, originally designed to map cyber-suite strategies and techniques, is quickly implemented for information operations, including the resolution campaign (Fernandez & Brezuk, 2024). To provide misinformation to attacks, security doctors can classify negative behavior in a structured way, allowing for more systematic detection and attribution. Disinformation strategies often rely on repeated dissemination to build perceived credibility, in a manner similar to money laundering, where information is circulated and amplified until it appears trustworthy. The MITRE ATT&CK framework helps organizations track and analyze these patterns across platforms. Accounting immunity and social techniques are as follows:

- **Disruptive actors often use legitimate news sites or influencers:** The rapid nature of MITRE ATT&CK provides a parallel to detect weight and an identification powder strategy to detect impersonation accounts (Government et al., 2022).
- **Creation and distribution through automation:** A technology such as Botnets or coordinated inhuman behavior falls within the category of MITRE ATT&CK, which is now being interpreted again for narrative spread (Sharma, 2022).

With the help of MITRE ATT&CK mapping, defenders can produce an opportunity for hazard resolution. For example, the knowledge of MITRE base analysts allows mapping bot-driven reinforcement for anti-techniques and estimating the possibility of the next stages. This structured approach allows industrial collaboration as organizations can detect dissolution strategies and share the unfortunate strategies behind them. In misunderstanding defense, MITRE ATT&CK acts not only as a classification of unfavorable behavior but also as a strategic planning tool for defenders, which allows systematic coverage of disintegration chains (Alevizos et al., 2022).

### 3.5.3 ISO/IEC 27001 and media integrity

While CSF and MITRE ATT&CK emphasize detection and response, ISO/IEC 27001 standards have a management-oriented approach, focusing on information protection and organizational control to maintain data integrity. Its relevance to misinformation restriction emphasizes the CIA triad of information funds (Din et al., 2024). Information protection for material authenticity, ISO/IEC 27001, requires organizations to install the information security management system (ISMS). Regarding incorrect information, the ISMS can be expanded to the material ecosystem, ensuring that only certified, authorized, and accepted materials are transferred to platforms.

**Some important practices are given below:**

- **Cryptographic attention of media:** Use encryption and blockchain-supported provenance posts to ensure that images, videos, and articles cannot be tampered with after publishing (Chen et al., 2023).
- **Access management:** Use identity management on content producers and ensure that only confirmed users can generate official material.
- **Economic alignment:** Built-in incorrect information limitation in standard operating procedures to manage the incident, thus normalizing material verification with cybersecurity breach (Government et al., 2022).
- **Audit and compliance:** Regular revision of digital media verifies the integrity of pipelines and ensures that the story is detected in history (Sharma, 2022).
- **ISO/IEC 27001:** also supplements zero confidence concepts by entering continuous verification in the government's disposition, ensuring that misinformation defense becomes part of organizational compliance culture.
- **Integrating frameworks for holistic defense:** While each framework offers a unique contribution to CSF in structuring defenses, MITRE ATT&CK adversary mapping, and ISO/IEC 27001 inside governance, integrating them creates a multilayered defense model to improve security. In this regard, the CSF protect function can lead to leveraging of ISO/IEC 27001 cryptographic attestation controls,

*Table 3.4* Summary of cybersecurity frameworks adapted for misinformation response

| *Reference* | *Cybersecurity framework* | *Core cybersecurity mechanisms* | *Adaptation for misinformation response* | *Technologies leveraged* |
|---|---|---|---|---|
| Fernandez et al. (2024) | ZTA critique and taxonomy | Identity-centric controls, policy enforcement, and trust zones | Establish granular trust boundaries for misinformation vectors.<br>Enforce dynamic access policies. | IAM, policy-as-code. |
| Chen et al. (2023) | ZTA for 6G networks | AI-driven trust evaluation, network slicing, and continuous authentication | Detect misinformation in ultra-low latency environments.<br>Isolate malicious slices. | AI analytics; 6G slicing. |
| Sarkar et al. (2022) | ZTA in cloud computing | Comparative analysis of ZTA models and threat surface reduction | Evaluate misinformation propagation across cloud tenants; apply least privilege. | Cloud-native ZTA and segmentation. |
| Sharma (2022) | ZTA implementation in the cloud | Micro-segmentation, continuous monitoring, and adaptive access | Real-time misinformation quarantine; forensic readiness in cloud logs | Telemetry and access control. |
| Alevizos et al. (2022) | Blockchain-enhanced ZTA endpoints | Decentralized trust, immutable logs, and endpoint verification | Validate misinformation sources via blockchain provenance.<br>Secure endpoint integrity. | Blockchain and endpoint ZTA. |
| Din (2024) | ZTA metaverse security | Blockchain-enabled identity; virtual trust zones | Mitigate misinformation in immersive environments; enforce avatar-level trust. | Blockchain and virtual identity. |
| Pooja et al. (2024) | ZTA for scientific collaboration | Secure data sharing and blockchain audit trails | Preventing misinformation in collaborative research.<br>Ensure data authenticity | Blockchain and ZTA sharing protocols. |

while MITRE ATT&CK mappings feed into CSF to detect anomalies. Furthermore, recovery strategies based on CSF might be aligned with ISO/IEC 27001 along with governance to ensure organizational learning as well as these frameworks that could be enhanced through blockchain as well as AI-enabled ZT models, offering proactive defense against traditional cyberattacks and emerging disinformation threats (Din et al., 2024). The cybersecurity framework summary is provided in Table 3.4.

## 3.6 INTEGRATED DEFENSE STRATEGIES: TOOLS AND TECHNOLOGIES

The rapid growth of AI-competent misinformation campaigns obliges robust defense mechanisms beyond ordinary identification and filtering. Recent defense robust strategies should be integrated into cybersecurity frameworks, ZTA, blockchain technologies, and AI-operated analysis to protect and defend against deepfake and widespread misinformation widespread. The current section highlights the integration of defense approaches through four domains: Far intelligence, Federated Identity and Access Management (FIAM), real-time content analysis, and secure blockchain.

### 3.6.1 Threat intelligence for misinformation sources

Regarding the core aspect in fighting misinformation is the ability to detect characteristics and reduce malicious power operations before they get traction. Threat selection has focused on identifying Indicators of Agreement (IOC) such as suspected domain registration, deviant IP activity, bottom coordination, and synthetic media proliferation (Mansur, 2025). These IOC are especially important for distinguishing legitimate stories from orchestrated suffering campaigns. Applying the reset principles shows no actor-incorporated or external-natural should be trusted (Fernandez & Brezuk, 2024). It never believes, always verifies "the attitude in accordance with intelligent ways that continuously validate data and its sources," and ensures that incorrect information is flagged in the network. At the national and geopolitical levels, misinformation is often aimed at important infrastructure and democratic institutions and requires intelligence-sharing ecosystems between governments, academics, and private organizations (Chuan et al., 2020). However, ZT implementation strategies could be applied for continuous monitoring in data pipelines, limiting the ability to insert misled data. Similarly, Huber and Kandah (2024) emphasized the role of zero confidence+ in securing the IoT on the scale of a rapidly relevant environment, where misinformation can be promoted through connected equipment and smart assistants. Emerging frameworks also emphasize the reliability of the AI system. Since the side effect utilizes Generic AI to

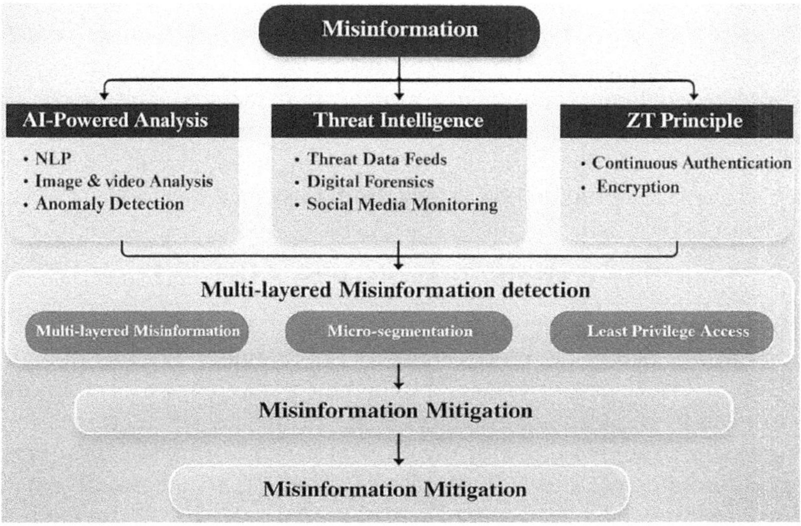


*Figure 3.8* Multilayered defense: combining AI, threat Intel, and ZT in misinformation detection.

reduce misunderstandings, there is a parallel need to secure intelligence in AI-driven danger intelligence platforms (Polemi et al., 2024) and ensure integrity. Therefore, integrated defense must balance intelligence on adversaries with safeguards against AI exploitation. Figure 3.8 demonstrates a multilayered defense that combines AI threat and ZT.

### 3.6.2 Federated identity and access management

One of the most important ambitions for misinformation campaigns is spreading false and synthetic identities on social platforms. FIAM introduces an overview of verifying digital identity across domains, ensuring that individuals who interact on platforms are certified by dependable suppliers. ZT and FIAM made a close difference. Iordache et al. (2022) argue that MFA and extended biometric verification significantly reduce the success of false account construction. It is especially relevant for social media platforms that encounter waves of synthetic identity used to reinforce fake news. The blockchain of identification verification provides an extra layer of competent decentralization protection. Alevizos et al. (2022) suggested taking advantage of blockchain to anchor claims of identity, which would make modeling or falsification quite an obstacle. Similarly, Pooja and Chandrakala (2024) reflect how blockchain-enabled ZT mechanism can enable the cooperation verification of identity in secure data sharing, and the research network approach applies equally to misinformation defense. From a comprehensive social point of view, FIAM can function as a preventive measure against large-scale incorrect information campaigns to increase the cost of building an account. However, Al-Tamimi and Abu Al-Hija (2024) ensure that these mechanisms are suitable for offering chain risk and technological advances, ensuring that identity control does not become obsolete because the opponents are evolving. Thus, FIAM, combined with ZT principles and blockchain verification, provides a powerful basis for protecting digital ecosystems from identified misunderstandings.

### 3.6.3 Real-time content analysis with AI/ML

The misinformation material requires real-time analysis operated by AI/ML. Natural Language Processing (NLP), computer vision, and audio analysis are synthetic stories, manipulated media, and key tools for coordinated promotion. With regard to NLP algorithms for the identification of synthetic text, one may look at the examination of semantic frameworks, relapse mechanisms, and stylometry abnormalities to determine whether content and activities are generated by AI (Mansurr, 2025). Advanced AI contemporarily requires real-time NLP to stop large-scale manipulations of text and other content. In sharp contrast to common knowledge concerning viral materials, NLP analyzes the laundering procedures for content and assesses how fake news circulates through various channels to gain legitimacy. The significance of deepfake technology and multimedia analysis cannot be overstated, however. Audio and video deepfakes are pivotal in the spread of disinformation (Chen et al., 2023), which lowers the trust threshold for the 4G network, where multimedia content will be predominant in communication. AI-assisted integration within forensic technology affords real-time deepfake detection through image analysis and synchronization evaluation, thereby providing users with adaptive AI protection. In spite of this, detection mechanisms still face unchallenged difficulties (Polemi et al., 2024). The focus is on the demands of technical precision so that AI can offer relief and provide flexibility against negative control. The incorporation of transparency within detection algorithms bolsters and enhances trust on the part of analysts.

Finally, the reset principles strengthen this defense by continuously verifying the current material. Huber and Kandah (2024) suggest use of ZT, where IoT is built directly into the IoT environment, ensuring that misinformation-laden media cannot propagate unchecked across devices.

### 3.6.4 Blockchain and secure content provenance

- **Construction of material origin and integrity:** Is one of the most important aspects of misinformation defense. Blockchain provides an irreversible account book that ensures that each piece of digital content, text, video, image, or data set can be traced back to the original manufacturer.

- **Unchanged material chain:** Pooja and Chandrakala (2024) discuss how the blockchain principles of ZTA support the secure review and sharing of scientific materials. This system uses the media and allows journalists, factors, facts, and platforms to work together to confirm authenticity without the risk of tampering.
- **Associated verification:** Pooja and Chandrakala (2024) explain how the blockchain-competent ZTA supports the secure review and sharing of scientific materials. This system uses the media and allows journalists, factors, facts, and platforms to work together to confirm authenticity without the risk of tampering.
- **National and institutional adoption:** Governments and public institutions are beginning to detect blockchain-based proven systems to defend themselves against foreign influence campaigns. Iordache et al. (2022) suggest that integration of increased biometric security tracking creates a double layer of defense, which determines both source identity and the media.
- **Supply chain security:** Al-Tamimi & Abu Al-Hija (2024) have expanded the approach to supply chains, where incorrect materials can infiltrate through compromised nodes. Blockchain provenance ensures end-to-end integrity in digital supply chains and corresponds to the principle of ZT to independently confirm each node. In this context, blockchain provides protection against misinformation and bolsters the essential trust foundation required for future digital ecosystems.

## 3.7 CASE STUDIES AND APPLICATIONS

In the ZTA case studies and applications section, the practical application integrated into defense strategies for misinformation and digital manipulation shows the effectiveness of ZTA, the blockchain system, and the hazard framework. This section illustrates the case studies depicting the distribution of these technologies in important domains such as social platforms, choices, and the government's verification.

### 3.7.1 Social platforms using ZTA for botnet mitigation

Social media sites regularly face the challenge of botnets that automate disinformation and inflate the volume of fake news and stories. To make ZT models more functional in the digital context, the general outer limits of the ZT model require systematic changes. The ZTA framework in NIST SP 800-207 allows platforms to avoid the constant verification of every user and transaction. In this situation, the health and behavioral properties of the unit are constantly evaluated before communication channels are unlocked. With regard to social networks, seeing real-time analysis as a measure of growth rather than an aggressive approach facilitates the detection and mitigation of botnet threats. One interesting application of blockchain technology is a facility for identifying qualified analysts of ZTA penetration, which integrates an anomaly detection algorithm with unchangeable audit logs. Using machines in conjunction with blockchain technology allows the platform to deliver tamper-proof surveillance while documenting attempted botnet activity in real-time. However, the problem of prions in ZTA-enabled social media platforms for botnets remains unsolved. There ZTA-enabled social media platforms need to control real-time responsiveness for withholding prions as well as user privacy and balance purpose, user count, and scalability.

### 3.7.2 Election disinformation monitoring with threat intelligence

Choices are the primary method used to evaluate conflict resolution campaigns, where those who are against voting try to alter the thoughts of others and make these individuals lose trust in democratic institutions by spreading misleading information and news. Using AI in deepfake is a source of threat intelligence that is very important for keeping an eye on such campaigns, concentrating on the agreement indications that come with making fake content, coordinated bot activity, and narrative reinforcement (Polemi et al., 2024). Election monitoring platforms quickly distributed ZTA principles in combination with blockchain to strengthen identity protection (Aleisa, 2025), showing blockchain-enabled ZTA's capacity to ensure authentication processes for privacy retention and privacy preservation during sensitive political events. This approach ensures that an actor analyst participating in voter monitoring platforms, journalists, analysts, and fact-checkers provides constant confirmation, reducing infiltration of malicious institutions. Furthermore, Awan et al. (2023) argues that the use of surveillance technologies with blockchains perhaps collectively shares implementation of access control mechanisms of the different authorities and private players to share. It practically guarantees that surveillance technologies will block surgery and identify and report unauthorized actor trespassers. These operational methodologies can be praised for the rapid responses. For instance, the NLP voting ML model for the scoped monitoring platform enables detection of fraudulent activities for the scheduled preliminary elections. In systems that contain blockchains, they provide certificates of confirmation for resolution campaigns, which can be altered through operational methodologies. These provide quick responses. However, most challenges lie on the scale of privacy and neutrality in politics (Rivera et al., 2024). Speaking of processes, elections should retain this balance since strong over-verification processes can block legitimate users, and on the other hand, weak verification systems can allow bad infiltration.

### 3.7.3 Government media verification system using blockchain

Governments may face various obstacles with verifying and communicating authentic information within saturated environments due to rapid synthetic media and misinformation, where the reliability in official communication is necessary for announcements on crises, elections, and politics and integrating to blockchain-based provenance system in the workflow for government media has emerged as a strong solution. Awan et al. (2023) explains how blockchain can be enhanced to create confidentiality frameworks to confirm the official media. By linking cryptographic evidence of authenticity for each press release, video, or declaration, governments can ensure that civil and international organizations can independently confirm the origin of the material (Rivera et al., 2024). Blockchain-based identity systems for MFA should be demonstrated to further support this approach, and only verified officers can make digital statements. Practical distribution suggests that the authorities can combine blockchain with AI-operated media forensics to detect manipulated images, videos, and audio (Alevizos et al., 2022; Kulkarni et al., 2024). One of the most significant benefits of the government's blockchain verification is civilian faith. By providing a transparent, proven investigation, the authorities strengthen individuals and organizations regardless of the validation of official media, reducing dependence on centralized gatekeepers, besides the implementation of challenges that include the cost of scaling such systems and ensuring dissimilarities with global information-sharing standards. Thus, the case study suggests that ZTA and blockchain integration may provide a robust basis for combating misinformation in various contexts. For social platforms, they provide effective defense against botnets and automated reinforcement. For choices, they enable privacy preservation and real-time monitoring through danger information. Also, for governments, they provide a verification-tested structure that strengthens civil confidence in the official media. Despite their promises, adoption challenges balance special secrecy with scalability, interoperability, and purpose. Constant progress in blockchain, AI/ML, and ZTA is required to limit these systems for extensive use in defense.

## 3.8 CHALLENGES AND FUTURE DIRECTIONS

AI competition-derogating misinformation presents challenges due to use of ZTA, which requires careful assessment to ensure safety and adaptation in a composite environment. One of the most important problems lies in the difference and integration complexity with cultural monuments. The infrastructure affected was not designed for the ZT principles, which complicates the surveillance mechanisms' uninterrupted access control, partition, and distribution. In Ahmed (2024), training and initiative struggle are observed when the existing organizational practice conflicts with ZTA management models. In addition, based on the cybersecurity policy structure in the government's settings, balanced differences with national security and privacy requirements require fine strategies (Abedi et al., 2024). Similar sellers and standards are required to integrate ZT into future generations of mobile networks (Cyber Security and Infrastructure Security Agency, 2023). Fragmentation with the CISA ZT maturity model presents complications (Ahn et al., 2024). Another challenge is privacy in data versus real-time monitoring. Effective information limitation often requires extensive telemetry collection, behavioral analysis, and relevant verification to validate the user, device, and content. However, real-time monitoring poses a sufficient risk to personal privacy, especially in an environment where individual and sensitive information is processed. Integrating ZTA with the MITRE ATT&CK framework contributes to cyber-resilience while ensuring a balance between visibility and privacy (Ali et al., 2022). Menard et al. (2025) proposed tailored maturity frameworks for multi-access edge computing that emphasize that data security necessitates adaptability in accordance with network-level monitoring. Thus, various research studies (Gurram, 2025) investigated data breaches. ZTA frameworks include monitoring and policies that must fit with countermeasures, notably MITRE ATT&CK, to prevent excessive data collection, where emerging issues resulting from AI models such as spoofing and adversarial evasion compromise the integrity of data (Joshi, 2024). Furthermore, with ZT framework deployments, agentic AI improves adaptive defenses and generates novel attack surfaces where generative adversarial techniques evade anomaly detection (Grandhi, 2025). Thus, it is emphasized that while AI drives the advancement of ZTA, resilience must be based on the adversarial robustness of the ML implementation systems. Although Identity and Access Management (IAM) is a fundamental component of ZTA, it emphasizes ML-driven risk assessment systems, which introduce novel vulnerabilities via adversarial spoofing (Yeoh et al., 2023; Z. ElSayed et al., 2024). However, the success of essential elements in ZT adoption is in protecting AI-driven decision systems against poisoning and evasion attacks. The urgency of this challenge is heightened in healthcare IoT, where ML approaches must endure adaptive adversaries (Laghari et al., 2025). The ultimate challenge is the establishment of universal content provenance standards to generate reliable digital ecosystems. In the absence of such measurements, misinformation efforts may persist in exploiting the disparity between detection and attribution (Al-Sharafi et al., 2023). Thus, it indicates that prospective of AI-driven ZT intrusion detection systems might address the vulnerabilities within industrial IoT, albeit their use is confined to certain sectors. IoT threat consumer detection methods with federated learning approaches emphasize that there is a need for provenance while maintaining distributed trust (Li et al., 2025). A revolutionary methodology for ZT foundation models (Bashir, 2024) contributes a collaborative AI framework that incorporates provenance, auditability, and trust verification inside the collaborative structure. Furthermore, Bashir (2024) asserts that ZTA could assist to improve cybersecurity independently while integrated with cohesive worldwide requirements that include traceability and verification era. These challenges underscore the need for cooperation among politicians, engineers, and AI researchers. ZT is also developed from technological improvements and a comprehensive framework that integrates interoperability, privateness, security, adverse resistance, and provenance to counter AI-pushed misinformation efforts.

## 3.9 CONCLUSION

To sum up, misinformation that generates employment AI and deepfake technology campaign poses one of the most formidable threats to modern integrity, the utilization of speed, scale, and personalization, which does not withstand traditional defense. This chapter highlights how ZTA provides a robust structure to minimize cyber risks that have occurred by misinformation through verifying constant confidence in users, equipment, and material sources. Recent research studies have attracted awareness of ZTA that provides a robust capacity. At the same time, distribution faces obstacles related to integration with cultural monuments, stress between privacy and monitoring, the absence of weaknesses, and universally demonstrated standards in the AI model, where cybersecurity structure plays an inevitable role in this context. ZT is a paradigm shift that redefines confidence as dynamic, relevant, and verifiable; it is more than simply a technological update. Regarding the role of the organizations, they can sufficiently limit the ability of malicious actors by implementing continuous authentication, real-time analysis, and unfavorable mighty AI models in the security architecture as well as latest update defense strategies. In addition, through references, this chapter highlights that cybersecurity models, federated learning, and identity management systems are converted to form a flexible backbone for future information integrity. Moreover, interdisciplinary collaboration, bridging technical, regulatory, and social domains, is needed going forward. Political decision makers should support interoperable and privacy-conservative surveillance standards, researchers must help develop a flexible AI model for detecting theft, and industry doctors should prefer proven verification as a shared responsibility. This integration of competence might be critical to prevent the information ecosystem from creating weapons against democratic institutions, economic stability, and social confidence. By leveraging intelligent cybersecurity frameworks, ZTA enables the society to establish an adaptive, transparent, and trustworthy information ecosystem, guaranteeing that the digital era is characterized by authenticity rather than manipulation and resilience.

## REFERENCES

Abedi, K., Nguyen, G. T., & Strufe, T. (2024). "Improving Resilience of Future Mobile Network Generations Implementing Zero Trust Paradigm," *NOMS 2024–2024 IEEE Network Operations and Management Symposium*, Seoul, Republic of Korea, pp. 1–5, https://doi.org/10.1109/NOMS59830.2024.10574898.

Ahmadi, S. (2024). Zero trust architecture in cloud networks: Application, challenges, and future opportunities. *Journal of Engineering Research and Reports*, *26*(2), 215–228.

Ahmed, F. (2024). Cybersecurity policy frameworks for AI in government: Balancing national security and privacy concerns. *International Journal of Multidisciplinary on Science and Management*, *1*(4), 43–53.

Ahn, G., Jang, J., Choi, S., & Shin, D. (2024). Research on improving cyber resilience by integrating the zero trust security model with the MITRE ATT&CK matrix. *IEEE Access*, *12*, 89291–89309.

Ajish, D. (2024). The significance of artificial intelligence in zero trust technologies: A comprehensive review. *Journal of Electrical Systems and Information Technology*, *11*, 30. https://doi.org/10.1186/s43067-024-00155-z.

Aleisa, M. A. (2025). Blockchain-enabled zero trust architecture for privacy-preserving cybersecurity in IoT environments. *IEEE Access*, *13*, 18660–18676. https://doi.org/10.1109/ACCESS.2025.3529309.

Alevizos, L., Ta, V. T., & Hashem Eiza, M. (2022). Augmenting zero trust architecture to endpoints using blockchain: A state-of-the-art review. *Security and Privacy*, *5*(1), e191.

Ali, B., Hijjawi, S., Campbell, L. H., Gregory, M. A., & Li, S. (2022). A maturity framework for zero-trust security in multi-access edge computing. *Security and Communication Networks*, *2022*(1), 3178760.

Altamimi, S., & Salameh, W. (2024). "Towards Analysis Detection of Deepfake Video via Deep Learning Models: A Review," *2024 International Jordanian Cybersecurity Conference (IJCC)*, Amman, Jordan, pp. 87–92, https://doi.org/10.1109/IJCC64742.2024.10847278.

Al-Sharafi, A. M., Alrayes, F. S., Alruwais, N., Maray, M., Alshuhail, A., & Darem, A. A. (2025). Ensuring zero trust security in consumer Internet of Things using federated learning-based attack detection model. *IEEE Access*, *13*, 54423–54438.

Al-Tamimi, S., Al-Haija, Q. A., & Alrawashdeh, K. (2024). "Zero-Trust Architecture for Securing Internet of Things (IoT) Networks: A Review," *2024 5th International Conference on Communications, Information, Electronic and Energy Systems (CIEES)*, Veliko Tarnovo, Bulgaria, pp. 1–6, https://doi.org/10.1109/CIEES62939.2024.10811176.

Al-Tamimi, S. A., & Abu Al-Haija, Q. (2024). Supply Chain Security, Technological Advancements, and Future Trends. In Q. Abu Al-Haija (Ed.), *Smart and Agile Cybersecurity for IoT and IIoT Environments* (pp. 211–234). IGI Global Scientific Publishing. https://doi.org/10.4018/979-8-3693-3451-5.ch010.

Awan, S. M., Azad, M. A., Arshad, J., Waheed, U., & Sharif, T. (2023). A blockchain-inspired attribute-based zero-trust access control model for IoT. *Information*, *14*(2), 129.

Bashir, T. (2024). Zero trust architecture: Enhancing cybersecurity in enterprise networks. *Journal of Computer Science and Technology Studies*, *6*(4), 54–59.

Brief, C. P. (2021). AI and the Future of Disinformation Campaigns. Center for Security and Emerging Technology, Georgetown University, Washington, DC, Technical Report.

Celeste, R., & Michael, S. (2021). Next-gen network security: Harnessing AI, zero trust, and cloud-native solutions to combat evolving cyber threats. *International Journal of Trend in Scientific Research and Development*, *5*(6), 2056–2069.

Chen, B., Qiao, S., Zhao, J., Liu, D., Shi, X., & Lyu, M. (2021). A security awareness and protection system for 5G smart healthcare based on zero-trust architecture. *IEEE Internet of Things Journal*, *8*(13), 10248–10263. https://doi.org/10.1109/JIOT.2020.3041042.

Chen, X., Feng, W., Ge, N., & Zhang, Y. (2023). Zero-trust architecture for 6G security. *IEEE Network*, *38*(4), 224–232.

Chuan, T., Lv, Y., Qi, Z., Xie, L., & Guo, W. (2020, November). An implementation method of zero-trust architecture. *Journal of Physics: Conference Series*, *1651*(1), 012010.

Cybersecurity and Infrastructure Security Agency. (2023). Zero trust maturity model. https://www.cisa.gov/sites/default/files/2023-04/zero_trust_maturity_model_v2_508.

Din, I. U., Khan, K. H., Almogren, A., Zareei, M., & Diaz, J. A. P. (2024). Securing the metaverse: A blockchain-enabled zero-trust architecture for virtual environments. *IEEE Access*, *12*, 92337–92347.

Edo, O. C., Tenebe, T., Etu, E. E., Ayuwu, A., Emakhu, J., & Adebiyi, S. (2022). Zero trust architecture: Trend and Impact on information security. *International Journal of Emerging Technology and Advanced Engineering*, *12*(7), 140.

Ejeofobiri, C. K., Adelere, M. A., & Shonubi, J. A. (2022). Developing adaptive cybersecurity architectures using zero trust models and AI-powered threat detection algorithms. *International Journal of Computer Applications Technology and Research*, *11*(12), 607–621.

ElSayed, Z., Elsayed, N., & Bay, S. (2024). "A Novel Zero-Trust Machine Learning Green Architecture for Healthcare IoT Cybersecurity: Review, Analysis, and Implementation," *SoutheastCon 2024*, Atlanta, GA, pp. 686–692.

Fernandez, E. B., & Brazhuk, A. (2024). A critical analysis of Zero Trust Architecture (ZTA). *Computer Standards & Interfaces*, *89*, 103832.

Grandhi, V. S. K. (2025). The role of identity and access management (IAM) in modern cybersecurity: Implementing zero trust principles for enhanced enterprise security. *World Journal of Advanced Engineering Technology and Sciences*, *15*(3), 179–186.

Gudala, L., Shaik, M., & Venkataramanan, S. (2021). Leveraging machine learning for enhanced threat detection and response in zero trust security frameworks: An exploration of real-time anomaly identification and adaptive mitigation strategies. *Journal of Artificial Intelligence Research*, *1*(2), 19–45.

Gurram, A. (2025). Generative AI for enhanced cybersecurity: Building a zero-trust architecture with agentic AI. *World Journal of Advanced Engineering Technology and Sciences*, *15*(1), 2380–2396.

Hattali, A. (2024). Zero-Trust Architectures in the Age of AI: Balancing Security and Efficiency in IT Systems.

He, Y., Huang, D., Chen, L., Ni, Y., & Ma, X. (2022). A survey on zero trust architecture: Challenges and future trends. *Wireless Communications and Mobile Computing*, *2022*(1), 6476274.

Huber, B., & Kandah, F. (2024). "Zero Trust+: A Trusted-based Zero Trust architecture for IoT at Scale," *2024 IEEE International Conference on Consumer Electronics (ICCE)*, Las Vegas, NV, pp. 1–6.

Iordache, C. A., Dragomir, A. V., & Marian, C. V. (2022). "Public Institutions Updated Enhanced Biometric Security, Zero Trust Architecture and Multi-Factor Authentication," *2022 International Symposium on Electronics and Telecommunications (ISETC)*, Timisoara, Romania, pp. 1–4, https://doi.org/10.1109/ISETC56213.2022.10010127.

Joshi, H. (2024). Emerging technologies driving zero trust maturity across industries. *IEEE Open Journal of the Computer Society*, *6*, 25–36. https://doi.org/10.1109/OJCS.2024.3505056.

Khan, M. J. (2023). Zero trust architecture: Redefining network security paradigms in the digital age. *World Journal of Advanced Research and Reviews*, *19*(3), 105–116.

Kolawole, I. (2025). Leveraging cloud-based AI and zero-trust architecture to enhance US cybersecurity and counteract foreign threats. *World Journal of Advanced Research and Reviews*, *25*(3), 006–025.

Kulkarni, A., Niraula, A., Bhattarai, H., & Niamat, M. (2024). "A Zero Trust Architecture employing Blockchain and Ring Oscillator Physical Unclonable Function for Internet-of-Things," *2024 IEEE International Conference on Electro Information Technology (eIT)*, Eau Claire, WI, pp. 508–513, https://doi.org/10.1109/eIT60633.2024.10609880.

Laghari, A. A., Khan, A. A., Ksibi, A., Hajjej, F., Kryvinska, N., Almadhor, A., ... & Alsubai, S. (2025). A novel and secure artificial intelligence-enabled zero-trust intrusion detection in an industrial Internet of Things architecture. *Scientific Reports*, *15*(1), 26843.

Li, K., Li, C., Yuan, X., Li, S., Zou, S., Ahmed, S. S., ... & Akan, O. B. (2025). Zero-trust foundation models: A new paradigm for secure and collaborative artificial intelligence for the Internet of Things. *arXiv preprint arXiv:2505.23792*.

Liu, C., Tan, R., Wu, Y., Feng, Y., Jin, Z., Zhang, F., Liu, Y., & Liu, Q. (2024). Dissecting zero trust: research landscape and its implementation in IoT. *Cybersecurity*, *7*, 20. https://doi.org/10.1186/s42400-024-00212-0.

Mansur, M. A. (2025). National security and cyber defense in the rise of artificial super intelligence. *ESI Preprints (European Scientific Journal, ESJ)*, *39*, 320–320.

Menard, P., Reyes, E., & Bateman, R. (2025). "Understanding Zero Trust Security Implementations via the MITRE ATT&CK and D3FEND Frameworks: Uncovering Trends Across a Decade of Breaches," *Proceedings of the 58th Hawaii International Conference on System Sciences*. Waikoloa Village: University of Hawaii.

Mensah, F. (2024). Zero trust architecture: A comprehensive review of principles, implementation strategies, and future directions in enterprise cybersecurity. *International Journal of Academic and Industrial Research Innovations (IJAIRI)*, *10*, 339–346.

Muriithi, G., Papari, B., Arsalan, A., Timilsina, L., Muriithi, A., & Buraimoh, E. (2025). Zero trust architecture for electric transportation systems: A systematic survey and deep learning framework for replay attack detection. *IEEE Open Journal of Vehicular Technology*, *6*, 2171–2194. https://doi.org/10.1109/OJVT.2025.3592041.

Muthusamy, K. (2025). Harnessing AI-powered zero trust architectures for proactive cyber defense: A comprehensive framework for future-ready network security ecosystems. *International Journal of AI, BigData, Computational and Management Studies*, *1*(1), 24–32.

Nahar, N., Andersson, K., Schelén, O., & Saguna, S. (2024). A survey on zero trust architecture: Applications and challenges of 6G networks. *IEEE Access*, *12*, 94753–94764.

Ofili, B. T., Erhabor, E. O., & Obasuyi, O. T. (2025). Enhancing federal cloud security with AI: Zero trust, threat intelligence, and CISA compliance. *World Journal of Advanced Research and Review*, *25*(2), 2377–2400.

Olanipekun, S. O. (2025). Computational propaganda and misinformation: AI technologies as tools of media manipulation. *World Journal of Advanced Research and Reviews*, *25*(1), 911–923.

Paul, E. M., Mmaduekwe, U., Kessie, J. D., & Dolapo, M. (2024). Zero trust architecture and AI: A synergistic approach to next-generation cybersecurity frameworks. *International Journal of Science and Research Archive*, *13*(2), 4159–4169.

Polemi, N., Praça, I., Kioskli, K., & Bécue, A. (2024). Challenges and efforts in managing AI trustworthiness risks: A state of knowledge. *Frontiers in Big Data*, *7*, 1381163.

Pooja, S., & Chandrakala, C. B. (2024). Secure reviewing and data sharing in scientific collaboration: Leveraging blockchain and zero trust architecture. *IEEE Access*, *12*, 92386–92399.

Rana, M. (2025). Enhancing zero trust cybersecurity with AI. *Journal of Information Systems Engineering & Management*, *10*(32s), 92–97. Available at SSRN 5216976.

Rivera, J. J. D., Muhammad, A., & Song, W. C. (2024). Securing digital identity in the zero trust architecture: A blockchain approach to privacy-focused multi-factor authentication. *IEEE Open Journal of the Communications Society*, *5*, 2792–2814. https://doi.org/10.1109/OJCOMS.2024.3391728

Romanishyn, A., Malytska, O., & Goncharuk, V. (2025). AI-driven disinformation: Policy recommendations for democratic resilience. *Frontiers in Artificial Intelligence*, *8*, 1569115.

Rose, S. W., Borchert, O., Mitchell, S., & Connelly, S. (2020). "Zero Trust Architecture," NIST Special Publication 800-207. https://doi.org/10.6028/NIST.SP.800-207.

Salem, A. H., Azzam, S. M., Emam, O. E., & Abohany, A. A. (2024). Advancing cybersecurity: A comprehensive review of AI-driven detection techniques. *J Big Data*, *11*, 105. https://doi.org/10.1186/s40537-024-00957-y.

Sarkar, S., Choudhary, G., Shandilya, S. K., Hussain, A., & Kim, H. (2022). Security of zero-trust networks in cloud computing: A comparative review. *Sustainability*, *14*(18), 11213.

Sharma, H. (2022). Zero trust in the cloud: Implementing zero trust architecture for enhanced cloud security. *ESP Journal of Engineering & Technology Advancements (ESPJETA)*, 2(2), 78–91.

Syed, N. F., Shah, S. W., Shaghaghi, A., Anwar, A., Baig, Z., & Doss, R. (2022). Zero trust architecture (ZTA): A comprehensive survey. *IEEE Access*, *10*, 57143–57179.

Tiwari, S., Sarma, W., & Srivastava, A. (2022). Integrating artificial intelligence with zero trust architecture: Enhancing adaptive security in the modern cyber threat landscape. *International Journal of Research and Analytical Reviews*, *9*, 712–728.

Ud Din, I., Habib Khan, K., Almogren, A., Zareei, M., & Arturo Pérez Díaz, J. (2024). Securing the metaverse: A blockchain-enabled zero-trust architecture for virtual environments. *IEEE Access*, *12*, 92337–92347. https://doi.org/10.1109/ACCESS.2024.3423400.

Yeoh, W., Liu, M., Shore, M., & Jiang, F. (2023). Zero trust cybersecurity: Critical success factors and A maturity assessment framework. *Computers & Security*, *133*, 103412.

Chapter 4

# AI-driven threats and countermeasures

## Securing the information space against synthetic media and deepfakes

*Qasem Abu Al-Haija*

### 4.1 INTRODUCTION

Generative Artificial Intelligence (GAI)'s dual-purpose abilities are based on extending model architectures and resources. Modern generative adversarial networks (GANs) and large language models (LLMs) can produce enough media with minimal user input, improving content creation and decreasing the risk of exploitation. According to Neupane et al. [1], threat actors progressively embrace such models to generate harmful narratives, manage social engineering attacks, and disrupt digital discourse. Obviously fabricated models such as FraudGPT, WormGPT, and custom DF engines seek to exploit vulnerabilities in public awareness, trust, and cybersecurity foundation [1].

The deepfake (DF) technique has developed rapidly and can now generate videos and voice recordings identical to authentic sources. Video DFs abuse GANs and neural networks to exchange faces and simulate realistic actions, while audio DFs mock vocal characteristics to personate individuals. Text generators such as GPT-4 can produce attractive news articles, interpretations, and social media posts that emerge as human-authored. Governments, financing, health, and public safety use synthetic media to reproduce falsehood.

Research studies exhibit the impact of synthetic information on people. A 2025 study concluded that DFs were fabricated due to a lack of trust in media corporations, representative discourse, and governance organizations. In early 2025, an analogous analysis signaled that GAI triggers division and permits rapid, large-scale misinformation (MI) while facing classical content verification systems. The United Nations' International Telecommunication Union (ITU) has sponsored organized measures and watermarking policies worldwide to contend with this issue.

Synthetic MI has a significant impact on various areas:

- Challenging experiences with DFs can reduce trust in news media, public figures, and even real audiovisual testimony, making persons more susceptible to manipulation or disengagement [2,3].

DOI: 10.1201/9781003626558-6

- Synthetic DFs directed at political managers or elections can control frustration or weaken democracy. The ITU's 2025 summit advocated stronger verification tools for defending GAI, which has been a tool for cybercrime [4].
- Victims of non-consensual DF pornography, voice scams mimicking family partners, or public assertions may have substantial emotional, reputational, and legitimate outcomes.

Chapter 4 provides a comprehensive analysis of the safeguarding techniques for the information space from threats from synthetic media. It examines the real-world environment and policy concerns. This chapter aims to teach researchers, technological experts, platform operatives, and policymakers about the various tools, frameworks, and methods that can assist in maintaining digital confidence in the era of GAI.

## 4.2 GAI AND THE MI ECOSYSTEM

The GAI revolution has created increasingly sophisticated synthetic media, DFs, AI-generated audio, text, and images, which can mimic reality with high fidelity. This section outlines the definition of DFs and synthetic media, including their types and enabling models/tools and the primary threat vectors and attack scenarios through which this content is weaponized.

### 4.2.1 DFs and synthetic media

DFs are synthetic media artifacts (audio, video, images, or text) that are generated or manipulated through machine learning (ML) techniques such as autoencoders, GANs, and LLMs to impersonate real individuals or contexts [2]. These include the following:

- **Video DFs:** Face-swapped videos and lip-synced distorted speech.
- **Audio DFs:** Voice cloning and fabricating speech using TTS systems.
- **Image synthesis:** Genuine photographs and scene handling
- **Text generation:** AI-based articles or social media posts that simulate human prose, e.g., GPT-based fake news.

The main GAI tools and models that enable MI include the following:

- **GAN-based systems**, such as StyleGAN and DeepFaceLab, are utilized for image and video synthesis [3].
- **LLMs** such as GPT-2, GPT-3.3, and GPT-4 can produce coherent but fabricated text content at scale [4].
- **Synthetic media platforms** like Synthesia for AI-driven video avatars and voice-cloning tools such as ElevenLabs, Speechify, PlayHT, and Descript are often used to impersonate individuals [2].

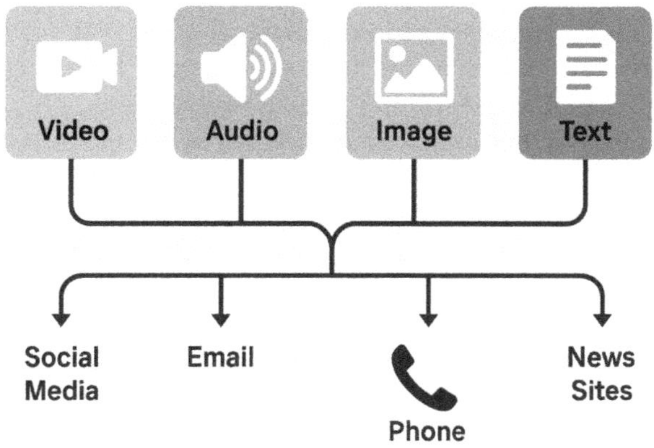


*Figure 4.1* Taxonomy of GAI-driven MI types and delivery vectors.

*Table 4.1* Contrasting common GAI models employed for MI

| *Model/tool* | *Modality* | *Typical misuse scenario* | *Ease of access* |
|---|---|---|---|
| StyleGAN, DeepFaceLab | Image/video | Face-swapped videos and DF clips | Open-source/free |
| GPT-22/GPT-3-3 GPT-4T-4 | Text | Automated fake news and social media posts | API-accessible |
| ElevenLabs, Descript | Audio | Voice cloning for fraud or impersonation | Web-based tools |
| Synthesia/AI avatars | Video | Propaganda avatars and political messaging | SaaS platforms |

This taxonomy is categorized in Figure 4.1, highlighting major GAI-driven MI types and delivery vectors, enabling readers to understand how different modalities map to threat situations.

### 4.2.2 Threat vectors and attack scenarios

**Political Manipulation:** DF content has been used in electoral settings. In the 2022–2024 period, DF political ads targeted candidates in the United States and other elections, including a viral AI-generated ad that led to the defeat of a Mayoral candidate in Louisiana [5]. Similar voice clones of British politicians raised concerns of interference in the UK general election [6].

**Identity Theft and Impersonation:** Audio DF scams have cost victims hundreds of billions; for instance, fraudsters have used real-time voice clones of CEOs or relatives to authorize transfers, thus exploiting voice authentication systems [7].

**Financial Fraud and Social Engineering:** GAI allows the creation of convincing fraudulent messages on an enormous scale. Fake insider articles, AI-written reviews, and chatbots such as WormGPT have enabled scams and phishing attacks with minimal human involvement.

**State-sponsored actors** and propaganda networks utilize synthetic video news anchors, AI-generated audio clips, and text outputs to influence public opinion. Taiwan saw political DFs attributed to disinformation arms intended to discredit public officials and manipulate voters [8].

These scenarios are categorized in Table 4.1, which compares the common generative models and tools employed in such MI activities, their modality, typical misuse, and ease of access.

## 4.3 ANATOMY OF DF CREATION

### 4.3.1 Technical pipeline

The standard workflow is depicted in Figure 4.2, indicating how DF media develops from data collection through synthesis and amalgamation, ensuing in a final polished performance.

- **Data collection** involves analyzing numerous data sources of individuals' images, video footage, and audio recordings. We often scrape and annotate public video clips, social media posts, and archival media to create a high-quality training organization. The DF synthesis relies on the quantity and diversity of reference data, making the final realism and fidelity of the synthetic media [9,10].
- **Model training** implies normalizing raw data (e.g., face alignment, cropping, frame selection, and to be used in ML models such as autoencoders, GANs, variational autoencoders (VAEs), or, more recently, diffusion models) [10]. Two AEs possess a common encoder for face swapping but have distinct decoders for source and target individualities, enabling the latent representation to map expressions between individuals [11]. A discriminator that constantly challenges the generator to create realistic images is introduced using GAN-based methods.

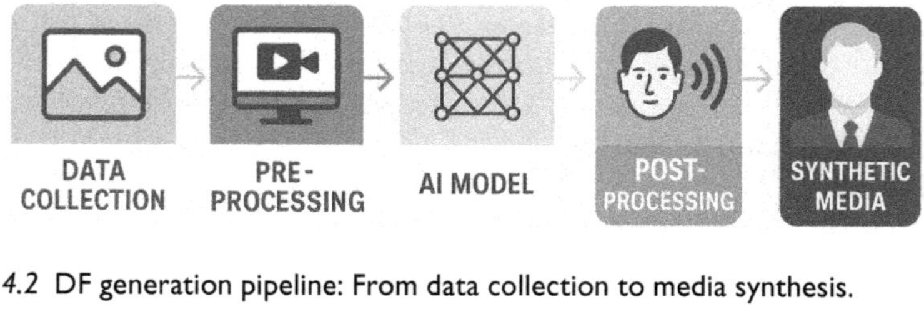


*Figure 4.2* DF generation pipeline: From data collection to media synthesis.

Recent advances in diffusion-based architecture have resulted in even greater fidelity outputs [9,10].

- **The post-processing phase** improves the raw generated output where synthesized faces are merged into target video frames, audio is labeled or manufactured via text-to-speech models, and color adjustment or noise mixing is utilized to remove visual artifacts. Audio DFs require voice cloning, prosody synchronization, and alignment with lip movements. The integration of visual and acoustic streams is intended to create a coherent multimodal output [9].

### 4.3.2 Tools and accessibility

The wide range of these tools enhances both legitimate and malicious use. Ease of access, combined with real-time generation pipelines, reduces the difficulty of access for MI campaigns.

- DeepFaceLab, FaceSwap, and other open-source frameworks provide complete face swapping and reenactment pipelines. DeepFaceLab is one of the most flexible and extensible platforms, providing face swaps with modular components for preprocessing, model training, and Blendi [11].
- Synthetic Media as a Service platforms such as Synthesia, Reface, or mobile apps such as Zao provide user-friendly interfaces where non-technical users can generate DF videos or audio without any effort. These services often utilize pre-trained models or remote inference Application Programming Interfaces (APIs), reducing the need for local GPU hardware and enabling a wide unauthorized use.

### 4.3.3 Detection challenges

The Detection Challenges collaborated with the Detection Challenges, where synthesis increases as detection increases, and vice versa.

- Newer GANs and diffusion-based syntheses produce images and videos with minimally visible images. Temporal consistency improvements (such as realistic eye blinking, natural facial expressions, and coherent speech-to-lip synchronization) lessen the efficacy of earlier detection heuristics based on intermittent or frame irregularity [10,12].
- As detection models improve, generative models increasingly utilize adversarial training to evade recognition. A classifier's decision can be reversed if small changes are present [9].

## 4.4 CYBERSECURITY APPROACHES TO DETECT AND COUNTER SYNTHETIC MEDIA

### 4.4.1 AI-based detection techniques

**Convolutional Neural Network** (CNN) models can identify distinct variations in facial expressions, blinking patterns, and head poses that are often overlooked by the human eye. Modern DF generators typically contain invisible spatial- or frequency-domain artifacts, such as warped shadows or unnatural skin reflections, that CNNs can detect. Figure 4.3 provides a typical CNN architecture for facial DF detection. The model incorporates several convolutional, pooling, and fully connected layers, using labeled datasets such as FaceForensics++ and Celeb-DF datasets. The state-of-the-art CNN-based detection networks, such as XceptionNet and EfficientNet, have achieved recognition accuracy, surpassing 90% on benchmark data, as well.

**Audio Forensics Using Spectrogram Analysis** is concerned with DF audio, which is produced by text-to-speech models and voice conversion systems. Scientists utilize spectrograms to identify frequency and amplitude patterns to identify synthetic audio. These spectrograms are then analyzed using CNNs or recurrent neural networks to identify temporal irregularities or fake harmonics. ASV spoof 2021, a challenge dataset for spoofed speech recognition, has enabled the expansion of deep learning models that identify interfering by evaluating time-frequency features among authentic and fake audio samples [13].

**Transformer-based models for Synthetic Text Classification** (TSTC) are concerned with AI-generated text, typically created using LLMs such as GPT-4; they can be challenging and straight-forward. To distinguish artificial text, TSTC systems such as RoBERTa and DeBERTa are fine-tuned to discover syntactical artifacts idiosyncratic to AI-made content. This includes exceedingly acknowledged phrasing, lack of genuine initiating, and stylometric consistency. Furthermore, hybrid approaches that combine

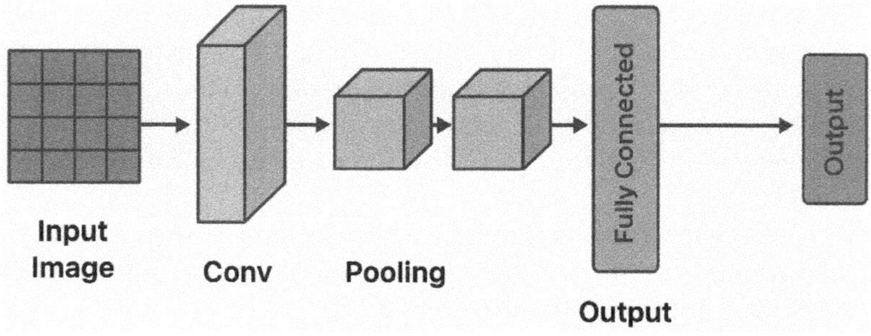


*Figure 4.3* CNN building blocks used for DF facial feature analysis.

stylometry and DL have demonstrated promise. Several systems require entropy-based randomness measures or examining sentence coherence to recognize synthetic writing [14].

### 4.4.2 Digital forensics and signal-based detection

Compression Patterns and Lighting Patterns can be identified as a further set of forensic characteristics by examining compression artifacts and lighting constancy. DF videos often have multiple instances of lossy compression, which can be detected by quantization matrix analysis. The use of mathematical methods can be advantageous in terms of lighting, shadows, specular highpoints, and reflections, allowing for a precise comparison to estimated physical lighting conditions. This approach enhances visual transparency and enhances a more engrossing and sensible experience. These signal-based methods often perform as a first-line or low-cost recognition method, particularly in resource-constrained areas.

### 4.4.3 Watermarking and media authentication

Watermarking is a classical method that is intended to be an initiative-taking counter measure against synthetic media. Passive watermarking results in revealing genuinely embedded patterns or camera sensor noise that are difficult to replicate for generative models. Furthermore, Photo-Response Non-Uniformity (PRNU) patterns, which are unique to camera sensors, can be extracted and compared to known devices for validation. In contrast, active watermarking produces robust and invisible codes directly into the media at the time of capture. These can include spatial frequency disturbances or color channel encodings that persist until the post-processing or compression process. Furthermore, more secure approaches include cryptographic watermarking, where a cryptographic hash of the content is signed using a private key and embedded into metadata. The hash is re-calculated and compared to the signed edition at the verification end, confirming integrity and authenticity. When linked with reliable execution positions or secure elements, cryptographic watermarking becomes tamper-evident and resistant to forgery. The Coalition for Content Provenance and Authenticity (C2PA) and Adobe's Content Credentials (C2) initiative are using these techniques in practical content workflows [15].

### 4.4.4 Blockchain-based content provenance

Blockchain technology (BT) is a decentralized and immutable approach to examine media origin and integrity throughout content life cycles. It is a source of information for the history of media. BT-enabled provenance systems have already been implemented. Starling Lab projects utilize BT and secure metadata to verify human rights documentation in courtrooms

*Table 4.2* Summary of detection techniques, accuracy levels, and computational costs

| *Detection method* | *Modality* | *Accuracy (%)* | *Real-time capability* | *Key limitations* |
|---|---|---|---|---|
| CNN-based visual detection | Video/image | 90–96 | High (GPU required) | Adversarial attack vulnerability |
| Audio spectrogram analysis | Audio | 85–92 | Moderate | Noise sensitivity |
| Transformer-based classifiers | Text | 80–90 | High | False positives in hybrid texts |
| PRNU and artifact forensics | Image/video | 70–85 | High | Sensitive to post-processing |
| BT provenance | All | N/A | Low | Storage and scalability overhead |

and public discourse. Furthermore, smart surveillance arrangements in law enforcement or critical infrastructure can monitor camera video recording and sensor data on-chain, facilitate secure playback audits, and enhance trust in high-risk ecosystems.

Figure 4.3 shows a DL model consisting of convolutional and pooling layers to identify and classify facial anomalies commonly encountered in AI-generated videos. Table 4.2 provides information on detection techniques, precision levels, and computational costs.

## 4.5 COUNTERMEASURES AND POLICY-LEVEL RESPONSES

### 4.5.1 Content authentication standards

The C2PA presents an open-source worldwide standard inventiveness that describes C2—a cryptographically signed metadata "label" attached to digital assets to verify origin, editing, and authenticity [16]. C2PA (developed by founding members such as Adobe, Microsoft, Google, Intel, Arm, and Truepic) allows embedded or sidecar exhibits to cover signed affirmations regarding derivation, content binding, and authenticity signatures. These standards provide a tamper-evident chain of provenance that is resilient to manipulation and effective in preventing AI-generated tampering in images, videos, or documents.

Originally an Adobe-led Content Authenticity Initiative (CAI), Project Origin relocated to C2PA, employing industry efforts with unified open standards. Adobe-C2 can monitor image and video creation workflows, flagging edits, and AI participation for downstream consumers [16]. In 2024–2025, YouTube and Google Search began incorporating C2PA metadata into their "about this image" or video-labeling features to highlight whether content was captured authentically or synthesized. However, a wider adoption remains uncertain; numerous smartphone vendors and major platforms have yet to provide interoperable metadata viewing or verification.

### 5.5.2 Platform-level defense mechanisms

At the platform level, real-time filtering and media flagging are essential to reduce the viral spread of synthetic MI. Platforms deploy automatic classifiers and label metadata (e.g., C2PA "made with AI" tags or proprietary AI-generation labels) to flag questionable content upon upload. In some services, user action, such as sharing, is delayed due to human moderation or algorithmic review. Community-based MI detection adds a crowdsourced layer: users report misleading synthetic media, and community moderators or external fact-checkers assess and tag the content. Platforms such as X and Facebook have experimented with internal and third-party fact-check partnerships to label manipulated media, though policies vary depending on the region and modality [16].

### 5.5.3 Legal and ethical implications

Synthetic Media's legal obligations are increasing to define and govern DFs. However, scholars contend that definitions within the AI Act remain vague, allowing ambiguity between permissible image editing and disallowed synthetic manipulation [17]. In 2023, the DFs Accountability Act (established in Congress in 2023) would require creators to disclose AI-altered content in the United States. This federal law complements laws in 21 states that criminalize non-consensual DF use and election-related manipulation.

Legislators are faced with an ethical challenge in combating malicious DFs while avoiding excessive censorship. Regulations must thoughtfully identify thresholds for prosecution (e.g., election timing limits and explicit non-consensual misuse) to prevent harming effects on lawful free expression or artistic parody. Transparency obligations under law must be based on contextual harms while safeguarding creative and journalistic freedoms.

Globally, policy trends suggest a growing interest in regulation. The EU AI Act is responsible for requiring transparency and labeling regulations for synthetic media. The US federal system lags but is accelerating due to the DFs Accountability Act. Other districts, such as China's Deep Synthesis Provisions and the UK's Online Safety Act (2023), also prohibit content disclosure or mandate removal of harmful DF content [18]. Coordinated

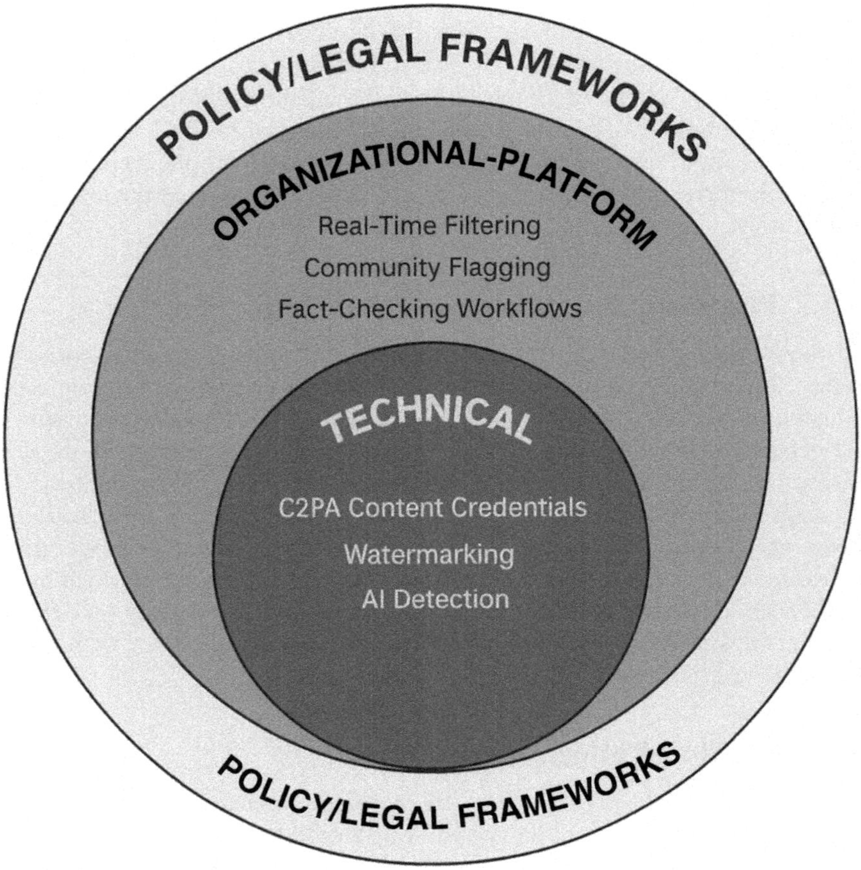


*Figure 4.4* Overview of technical, organizational, and policy-level defense layers.

global standards will be essential to preventing cross-border DF propagation and preserving democratic standards.

Figure 4.4 presents an integrated visualization of how standards, platforms, and regulations interact to form a comprehensive defense strategy against synthetic MI. It exemplifies three concentrical layers: technical (C2PA Content Credentials, watermarking, and AI detection); organizational/platform (real-time filtering, community flagging, and fact-checking workflows); policy/legal frameworks (AI Act, DFs Accountability Act, and international transparency mandates).

## 4.6 CASE STUDIES AND APPLICATIONS

This section examines three real-world incidents illustrating how GAI-driven MI has been deployed, as well as corresponding detection and mitigation strategies.

*Table 4.3* Comparative evaluation of real-world MI experiences and counter measures

| *Case* | *Modality* | *Threat tactics* | *Detection and Forensics methods* | *Outcome/ mitigation* |
|---|---|---|---|---|
| 2020 U.S. Election | Video/ audio | Bot-amplified rumors and misleading edits | Manual fact-check, reverse search, and early neural detection | Limited reach, effective public warnings, and takedowns |
| Celebrity Voice Scams | Audio | AI voice cloning and impersonation | Spectrogram analysis, speaker verification, and anti-spoof models | Financial losses, warnings, and MFA mitigations enacted |
| Ukraine–Russia war | Video (DF) | Fake surrender clips and propaganda | Frame artifact analysis, metadata, and semantic integrity models | Videos are debunked and removed rapidly by platforms |

### 4.6.1 2020 U.S. election DF incidents

Despite widespread fear in the aftermath of the 2020 U.S. election, the actual impact of DFs still remained limited. Experts had warned that manipulated video or audio could alter public perception by portraying political figures making false statements or engaging in controversial behavior. In practice, most viral MI was either low-tech manipulation (e.g., slowing of the video of Speaker Nancy Pelosi or rumor amplification—not fully AI-generated DFs [19]. The perpetrators attempted to uncover false information through social media platforms, influencer accounts, and fringe forums [19]. Human review often employed automated methods to identify manipulations in the context [20].

### 4.6.2 Media forensics in Ukraine–Russia War 2022

A remarkable illustration was a video of President Zelenskyy released by Russian media urging Ukrainians to surrender. The video had briefly dispersed online before Facebook, YouTube, and other platforms quickly debunked and removed it (Table 4.3) [21].

## 4.7 CHALLENGES AND FUTURE DIRECTIONS

### 4.7.1 Detection arms race through adversarial learning and spoofing models

In an escalating arms race, the DF generation and detection are locked. As detectors become more advanced, adversaries employ adversarial repercussions and training-aware spoofing to prevent them from escaping.

Adversarial Feature Similarity Learning has been anticipated to increase detector power by imposing feature similarity in class and variation across real and fake media. This has improved the effectiveness of adversarial training on datasets like FaceForensics++ and DeeperForensics [22].

### 4.7.2 Need for standardization: common APIs and evaluation benchmarks

Discrepancy in datasets, metrics, and experimental configurations has facilitated a fair comparison of detector models. The DF Bench model [23] proposed a comprehensive benchmarking to enhance reproducibility and transparency. The model utilized multiple state-of-the-art methods, datasets, and standardized evaluation metrics and protocols. Also, Firc et al. [24] developed a comprehensive evaluation framework for speech DF detection, evaluating 40 models under various conditions to enhance the comparability of techniques. However, APIs, versioned evaluation codebases, and dynamic challenge platforms remain restricted.

### 4.7.3 Explainability and trust: black box models vs. interpretable detection

Modern detection models are based heavily on deep neural networks that commonly employ opaque, black box classifications. A topical technique utilizes Explainable AI to differentiate adversarial attacks against DF detectors. They utilize interpretability maps and incorporate them along inputs to enhance classification dependability, without altering detector performance [25]. Identical features such as eyeblink anomalies or spectrogram artifact patterns can also enhance trust by pointing to forensic artifacts, but their generalization and consistency vary across datasets. Explainability frameworks must become essential for detection pipelines, not an afterthought.

### 4.7.4 Emerging trends

Standard supervised detectors do not comprehend generative architectures. Zero-shot (ZS) techniques have been developed to identify DFs without prior training techniques. This can occur by learning general anomaly detection and representation embedding functions. For instance, Babaei and Banitalebi-Dehkordi [26] identify ZS learning frameworks that recognize formerly unseen DF types by utilizing generative model structure and distributional outlier detection. This capability is increasingly important as GAI evolves faster than retraining cycles.

Additionally, single-modality detectors for image- or audio-only outputs miss cross-modal interruptions. A comprehensive survey by Nguyen-Le et al. shows that multimodal detection (e.g., synchrony between lip movements and audio and cross-validation of visual and textual clues) increases robustness and generalization [27]. Fusion methods using attention, contrastive learning, and joint embeddings are effective for talking face- to-face DFs, mixed audiovisual scams, and synthetic-plus-image content. It is necessary to integrate multimodal detection as real-time, resilient modules.

## REFERENCES


1. S. Neupane et al., "Impacts and Risk of Generative AI Technology on Cyber Defense,"*arXiv*, Jun. 22, 2023. [Online]. Available: https://arxiv.org/abs/2306.13033
2. P. L. Kharvi, "Understanding the Impact of AI-Generated Deepfakes on Public Opinion, Political Discourse, and Personal Security in Social Media," *Computer*, vol. 58, no. 4, pp. 115–122, Apr. 2024. [Online]. Available: https://www.computer.org/csdl/magazine/sp/2024/04/10552098/1XApkaTs5l6
3. T. M. Abraham, "Leveraging data analytics for detection and impact of fake news and deepfakes," *Humanities and Social Sciences Communications*, vol. 12, Article 1040, 2025. [Online]. Available: https://www.nature.com/articles/s41599-025-05389-4
4. O. Le Poidevin. "UN Report Urges Stronger Measures to Detect AI-driven Deepfakes," *Reuters*, Jul. 11, 2025. [Online]. Available: https://www.reuters.com/business/un-report-urges-stronger-measures-detect-ai-driven-deepfakes-2025-07-11/
5. F.-A. Croitoru et al., "Deepfake Media Generation and Detection in the Generative AI Era: A Survey and Outlook," *arXiv preprint arXiv:2411.19537*, Nov. 2024. [Online]. Available: https://arxiv.org/abs/2411.19537
6. N. A. Chandra et al., "Deepfake Eval 2024: A Multi-Moda In-the-Wild Benchmark of Deepfakes Circulated in 2024,"*arXiv preprint arXiv:2503.02857*, Mar. 2025. [Online]. Available: https://arxiv.org/abs/2503.02857
7. S. van der Linden, "AI-Generated Fake News Is Coming to an Election Near You," *Wired*, Jan. 22, 2024. [Online]. Available: https://www.wired.com/story/ai-generated-fake-news-is-coming-to-an-election-near-you
8. A. Yiğitalp Tulga, "The Malicious Exploitation of Deepfake Technology: Political Manipulation, Disinformation, and Privacy Violations in Taiwan," *Global Taiwan*, May 7, 2025. [Online]. Available: https://globaltaiwan.org/2025/05/the-malicious-exploitation-of-deepfake-technology/
9. G. Pei et al., "Deepfake Generation and Detection: A Benchmark and Survey," *arXiv*, Mar. 2024. [Online]. Available: https://arxiv.org/abs/2403.17881
10. F. Abbas et al., "A Systematic Review of Deepfake Detection and Generation" *Expert Systems with Applications*, vol. 252, Article 124260, 2024. [Online]. Available: https://www.sciencedirect.com/science/article/pii/S0957417424011266
11. I. Perov et al., "DeepFaceLab: Integrated, Flexible, and Extensible Face-swapping Framework," *arXiv*, May 2020. [Online]. Available: https://arxiv.org/abs/2005.05535
12. N. Khan et al., "Unmasking Synthetic Realities in Generative AI: A Comprehensive Review of Adversarially Robust Deepfake Detection Systems," *arXiv*, Jul. 2025. [Online]. Available: https://arxiv.org/abs/2507.21157

13. X. Wang et al., "ASVspoof 2021: Automatic Speaker Verification Spoofing and Countermeasures Challenge Evaluation Plan," arXiv preprint arXiv: 2109.00537, 2021. https://arxiv.org/abs/2109.00537
14. R. Krishna et al., "Detecting AI-Generated Text Using Stylometric Analysis and Deep Learning," *ACM Transactions on Information Systems (TOIS)*, vol. 40, no. 3, Article 56, May 2022. https://doi.org/10.1145/3490481
15. E. Bureacă and I. Aciobăniței, "A Blockchain based Framework for Content Provenance and Authenticity," *2024 16th International Conference on Electronics, Computers and Artificial Intelligence (ECAI)*, Iasi, Romania, 2024, pp. 1–5, doi: 10.1109/ECAI61503.2024.10607477
16. J. C. Simmons and J. M. Winograd, "Interoperable Provenance Authentication of Broadcast Media Using Open Standards-based Metadata, Watermarking and Cryptography," *arXiv*, May 2024. [Online]. Available: https://arxiv.org/abs/2405.12336
17. B. Rijsbosch, G. van Dijck, and K. Kollnig, "Adoption of Watermarking for Generative AI Systems in Practice and Implications Under the New EU AI Act," *arXiv*, Mar. 23, 2025. [Online]. Available: https://arxiv.org/abs/2503.18156
18. K. Meding and C. Sorge, "What Constitutes a Deep Fake? The Blurry Line Between Legitimate Processing and Manipulation Under the EU AI Act." *CSLAW '25: Symposium on Computer Science and Law*, arXiv, Munich Germany March 25–27, 2025. [Online]. Available: https://arxiv.org/abs/2412.09961
19. J. Villasenor, "Deepfakes, Social Media, and the 2020 Election," The Brookings Institution, Jun. 2019. [Online]. Available: https://www.brookings.edu/articles/deepfakes-social-media-and-the-2020-election/
20. Wikipedia, "Audio Deepfake," 2025. [Online]. Available: https://en.wikipedia.org/wiki/Audio_deepfake
21. S. M. Qureshi et al., "Deepfake Forensics: A Survey of Digital Forensic Methods for Media Authenticity Analysis," *PLoS One*, vol. 19, no. 2, p. e0282039, Feb. 2024. [Online]. Available: https://www.ncbi.nlm.nih.gov/pmc/articles/PMC11157519/
22. S. Khan, "Adversarially Robust Deepfake Detection via Adversarial Feature Similarity Learning," arXiv preprint, Feb. 2024. [Online]. Available: https://arxiv.org/abs/2403.08806
23. Z. Yan et al., "DeepfakeBench: A Comprehensive Benchmark of Deepfake Detection," arXiv preprint, Jul. 2023. [Online]. Available: https://arxiv.org/abs/2307.01426
24. A. Firc et al., "Evaluation Framework for Deepfake Speech Detection," *Cybersecurity (SpringerOpen)*, vol. 8, no. 1, Article 50, 2025. [Online]. Available: https://cybersecurity.springeropen.com/articles/10.1186/s42400-024-00346-1
25. B. Pinhasov et al., "XAI-Based Detection of Adversarial Attacks on Deepfake Detectors," arXiv preprint, Mar. 2024. [Online]. Available: https://arxiv.org/abs/2403.02955
26. M. Babaei and S. Banitalebi-Dehkordi, "A Zero-Shot Learning Framework for Deepfake Video Detection," *Information*, vol. 14, no. 1, 2025. [Online]. Available: https://www.themoonlight.io/en/review/passive-deepfake-detection-across-multi-modalities-a-comprehensive-survey
27. H.-H. Nguyen-Le et al., "Passive Deepfake Detection across Multi-modalities: A Comprehensive Survey," arXiv preprint, Nov. 2024. [Online]. Available: https://arxiv.org/abs/2411.17911

Chapter 5

# Decentralizing truth

## A blockchain-based framework for trustworthy information in the GenAI era

*Meysam Abedi, Ismaila Temitayo Sanusi, and Markku Tukiainen*

## 5.1 INTRODUCTION

### 5.1.1 The trust crisis in our digital world

The way we create, share, and take in information has been totally flipped upside down by the digital age. Sure, it's made us more connected and has opened up knowledge to pretty much everyone, but it's also sparked a massive trust meltdown. Reports like the Edelman Trust Barometer (2024) reveal that in a lot of countries, less than half of the people actually believe what traditional media is putting out there. This mess stems from a few key issues that feed into each other: unverified stuff going viral, disinformation efforts getting smarter and sneakier, and old-school verification methods just not keeping up with tech that's evolving at breakneck speed. The fallout is huge—take the 2024 U.S. elections, where made-up stories threw everything into chaos, or the COVID-19 era, when bogus info messed with healthcare initiatives and saved lives (Kitchin, 2014; Fang et al., 2024).

### 5.1.2 Generative AI: opportunities and risks

With the growth of generative artificial intelligence (GenAI), everything has grown, and some things have even gotten worse. These cutting-edge models can create text, images, and sounds that are highly similar to humans (Alwaisi et al., 2023). On the positive side, they open all sorts of creative possibilities, but they also expose the reliability of information to a serious threat and turn accountability into hard nuts to crack (Calzada et al., 2025). This paradox is strange: technology, which is very good at lying, can also provide power tools to identify them. For example, something like GPT-5 can publish news pieces faster and on a larger scale than we have ever imagined (Park et al., 2025). But here's the rub—it's lopsided. People who study the facts may take a few minutes per case, while AI discloses fakes in a blink. The European Commission's insistence on "reliable artificial intelligence" really makes you think about it: for whom, exactly, and in what scenarios? (European Commission, 2024; European Parliament and Council, 2024).

DOI: 10.1201/9781003626558-7 

### 5.1.3 Limitations of centralized systems and research gap

Current confirmation tools, such as reality control organizations, state agencies, and social media, rely on centralized structures that show considerable limits. These systems are vulnerable to individual failures, lack transparency in decision-making, and are unable to scale effectively to deal with a large volume of content produced by AI (Wu & Seneviratne, 2025). Moreover, they are susceptible to political or financial influence, which undermines public confidence. Blockchain technology offers a promising alternative with its unchanging record and visibility (Politou et al., 2019). However, existing research has not adequately examined the structure of AI, and also to deal with inaccurate information. Most blockchain research has stuck to areas like confirming academic qualifications or keeping supply chains honest (Queiroz et al., 2019; Alshahrani et al., 2022), and AI-focused work has mostly looked at the dangers and ways to detect them (Weidinger et al., 2023; Allen & Weyl, 2024). There is a clear need for a comprehensive framework that unites these approaches. This chapter deals with the gap by proposing a hybrid architecture that combines the origin of blockchain content with the approval of AI and provides both technical and social accuracy.

### 5.1.4 Structure of this chapter

The remaining part of this chapter is structured as follows: Section 5.2 addresses how the incorrect information in the world of GenAI changes and explains the key part of blockchain in the development of trust. Section 5.3 describes the fundamentals of our proposed framework and main components. Section 5.4 provides an overview of how it all works in practice, with a focus on weaving AI in as a helper for checks. Section 5.5 covers the real-world, moral aspect, and obstacles of governance. Finally, Section 5.6 addresses scalability and future developments, such as quantum cryptography and broader social concepts.

## 5.2 LITERATURE REVIEW AND RESEARCH POSITIONING

### 5.2.1 The shifting world of misinformation in the age of generative AI

The rapid expansion of digital media and social networks has fundamentally changed the mechanisms of information access and dissemination. This spread of access to knowledge has created an environment highly susceptible to the dissemination of false information and lies (Kitchin, 2014).

Empirical studies show that emotionally charged falsehoods, in particular, tend to achieve greater access and speed than actual information (Fang et al., 2024).

The emergence of GenAI tools has made the landscape more complex. Advanced systems, such as GPT-5 and deepfake generators, now produce text, images, and voice that are increasingly undetectable from the valid human output (Europol, 2022). The risks of this misuse are particularly acute in high-stakes domains such as healthcare, law enforcement, and governance, where sophisticated AI-generated content could erode democratic principles, spread disinformation, or reinforce societal biases (Yang et al., 2024; Insight & Foresight, 2024; Amoore et al., 2024; Chafetz et al., 2024; Delacroix, 2024; Gabriel et al., 2024; Shin et al., 2024; Tsai et al., 2024; Weidinger et al., 2023; Allen & Weyl, 2024; Cugurullo et al., 2024; Farina et al., 2023; Calzada, 2021; Aguerre et al., 2024; Angelidou & Sofianos, 2024).

Currently, the dominant strategies to combat incorrect information are centralized and rely on reality control organizations and the original social media. However, these centralized models reflect an important vulnerability: they are prone to single failure points, have a lack of clarity in their decision-making processes, and are unable to scale effectively against the large volume of produced content by AI. These limitations are in dire need of a decentralized and flexible approach to build trust in digital information ecosystems.

### 5.2.2 Blockchain as a foundation for trust

Blockchain stands out as a solid counter to those centralized systems, thanks to its focus on spreading out control, making records unchangeable, and keeping everything transparent (Politou et al., 2019). Once something's logged on a blockchain, tampering with it raises alarms—it's that tamper-proof (England et al., 2020). This makes it perfect for tracking where content comes from and how it's evolved, creating a reliable trail. In addition, by not having one boss in charge, it shares trust around the network and cuts down the risk of single points of failure or centralized manipulation (Mafrur, 2025). Previous research has examined the application of blockchain technology in domains such as academic credential verification (Alshahrani et al., 2022) and supply chain management (Queiroz et al., 2019). However, the integration of blockchain with AI to address the unique challenges posed by GenAI-driven misinformation remains insufficiently explored. This chapter seeks to address this research gap by advancing the current discourse and proposing a comprehensive hybrid framework. This chapter focuses on expanding on what's out there.

The proposed contribution is a novel hybrid architecture that leverages the immutability of blockchain to establish content provenance, while incorporating AI to enhance verification processes. In contrast to previous approaches that have focused exclusively on either blockchain or AI-based detection, this integrated framework is specifically designed to address the scale and velocity of challenges introduced by contemporary GenAI technologies.

## 5.3 PROPOSED FRAMEWORK DESIGN AND CORE COMPONENTS

By delving into the crisis of information in the GenAI era and the pivotal role the blockchain can play in addressing it, this chapter focuses on the actual design of the proposed solution. The purpose of this project is to build a system that is designed not only to be decentralized and transparent but also to be flexible and scalable at the ground level to withstand the ever-increasing amount of inaccurate digital information. What follows is the discovery of the main ideas that form architecture, the logic behind the choices of technology, and a closer look at the underlying components that bring the frame to life.

### 5.3.1 Design principles for a blockchain-based trust framework

The development of a flexible framework-based trust requires adherence to a set of core design principles that ensure its effectiveness, stability, and widespread acceptance. These serve as the primary design principles for the architecture, integrating different technological components and defining the operational philosophy of the system.

First, transparency is a priority. Any important action within the framework, from the creation of content and the initial approval to further corrections and resolution, should be registered in the blockchain. This ensures that a complete and approved audit trail is available to all authorized participants and that they can independently verify the origin of the information and the date of trust. This principle directly addresses the inherent opacity of the centralized operating system, where the median of content and algorithmic biases is often hidden.

Second, decentralization and resilience are critical to eliminate single points of failure and resist censorship. This framework should operate on a distributed network where no entity has any additional control over data or processes. This not only increases security against malicious attacks or systemic failures but also ensures that the system remains operational even if parts of the network are exposed to danger. With the distribution of trust over multiple nodes, the framework avoids the vulnerability associated with decentralized organizations and encourages a stronger and democratic information ecosystem.

Third, enabling and controlling the user are essential. This framework should enable individuals to control their digital identity and the origin of the content they create or interact with. This includes giving users the means to sign their contents, verify information, and participate in community-based certification processes. This transfers power from the media to the individual and reinforces a sense of ownership and responsibility in the information perspective.

Finally, interoperability and scalability are essential for practical implementation and future growth. This framework should be designed for integrating platforms and future technologies to prevent the creation of discrete data silos. In addition, it should be able to continuously manage a growing volume of content and user interactions without compromising performance or security. This requires careful attention to the underlying technologies and their capacity to effectively process transactions and maintain decentralization.

### 5.3.2 Selection criteria for underlying technologies

The effectiveness of any blockchain-based framework hinges significantly on the judicious selection of its underlying technologies. This process is directed by a comprehensive set of criteria to ensure alignment of selected components with the principles of the design of the framework and operational requirements.

The main criterion is security and invariance. The selected blockchain platform should provide strong encryption to ensure data integrity and prevent manipulation. This algorithm consists of strong hash algorithms, secure agreement mechanisms, and joint attack vectors. The invariance of the recorded data is not negotiable and provides the basic trust layer for the content source.

Decentralization and a consensus mechanism are equally critical. The selected technology should support decentralized network architecture and avoid relying on a small number of centralized or centralized control points. The consensus mechanism (e.g., Proof of Stake, Proof of Work, or delegated variants) should be evaluated for its ability to maintain network integration, resist spot fixing, and ensure fair participation.

Scalability and operational power are practical considerations for accepting the real world. Given the enormous volume of digital content, the selected blockchain should be able to handle a large number of transactions per second (TPS) without imposing high costs or significant delays. While pure focus is often accompanied by trade scaling, the framework seeks a balanced approach, with potential leveraging Layer 2 solutions or sharding techniques.

The interoperability and maturity of the ecosystem refer to the ability of selected technologies to communicate with other systems and their constructive society. A mature ecosystem with well-documented Application

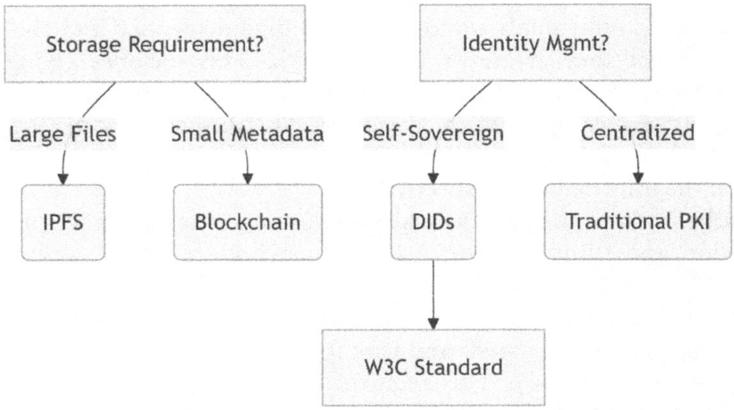


*Figure 5.1* Decision flow for selecting blockchain components.

Programming Interfaces (APIs) and a wide range of available tools facilitates more integration and future development. The bridge's capacity with other blockchains and traditional databases is also a significant advantage.

Finally, cost-effectiveness and energy efficiency are increasingly important. Transaction costs and the environmental impact of the consensus mechanism must be considered for long-term sustainability. Technologies that provide lower exchange costs and a more sustainable energy footprint are preferred, especially for public-use infrastructure, such as an information support framework.

Figure 5.1 provides a comprehensive decision-making tree for the selection of blockchain components, which, in particular, address the dual requirements of the framework: (1) decentralized storage solutions to maintain content and (2) authentication systems for verifiable authorship. The workflow systematically evaluates each technology against security measures, scalability, and interoperability and ultimately guides the selection of IPFS for storage and decentralized identifiers (DIDs) for identity management.

### 5.3.3 Core components: decentralized identifiers (DIDs), smart contracts, and IPFS

The proposed framework is based on blockchain for reliable information about three major technological underpinnings: DIDs, smart contracts, and the Interplanetary File System (IPFS). Each component plays a distinct and continuous role in creating the content's origin, automating authentication processes, and ensuring decentralized storage.

DIDs represent a paradigm shift in digital identity management, moving from centralized authorities to independent control (Voshmgir, 2020). DIDs are unique and permanent identifiers that can be acknowledged and do not need to be registered. Within this framework, DIDs enable content builders, publishers, and other parties to create a reliable digital identity without relying on traditional identity providers. When a piece of content is created, the creator's DID can be cryptographically linked to it, providing an undeniable and auditable proof of authorship. This mechanism is very important for combating content produced by AI because it allows clear documentation of the artificial media to its source, irrespective of whether it is an AI model or what it establishes. DIDs enhance accountability and strengthen the trust chain from the source.

Intelligent contracts—self-fulfilling contracts with agreement conditions directly embedded in the written code—constitute the backbone of automated verification and framework governance mechanisms. These contracts, deployed on a blockchain (such as Ethereum), automatically perform predefined actions when specific conditions are met and remove the need for mediators (Liang, 2025). Within the framework, smart contracts can be programmed to the following:

- **Automate content hashing and anchoring:** Automatically generate the content of content encryption and register them in blockchain after submission.
- **Manage verification workflows:** Facilitating and recording the community or expert approval process, including voting mechanisms and credit scoring for verifiers.
- **Enforce content policies:** Automatically flag or quarantine content that violates predefined rules, such as detected deepfakes or content from known malicious DIDs.
- **Incentivize participation:** Reward distribution or fine management for participants based on their participation in their approval and accuracy of their evaluations. This automation ensures efficiency, transparency, and impartiality in the authentication process and makes the system scalable and resistant to human bias.

The IPFS addresses the challenge of storing large digital content files in a decentralized and efficient manner. Blockchains are not designed to store large amounts of data due to their scalability and high costs. The IPFS provides a distributed file system in which the content is addressed instead of its location (such as traditional URLs) by the encryption combination. In the proposed framework, the actual content (e.g., a high-resolution image, a video, or a long text document) is stored in the IPFS, while only the unique encryption combination is recorded in blockchain (Saif et al., 2024). This approach has several advantages:

- **Decentralized storage:** the content is distributed in several nodes, ensuring resilience and preventing individual points of failure for storing.
- **Content addressing:** Content is retrieved based on its hash, meaning that as long as the content exists anywhere in the IPFS network, it

*Figure 5.2* Architecture of the decentralized verification framework.

can be accessed. This also ensures that if the content changes, its hash changes, and the creation of a link on an invalid blockchain is then immediately detected, thus signaling tampering.

- **Efficiency**: It avoids bloating the blockchain with large files, keeping transaction costs low and network performance high. This combination of blockchain for immutable metadata and IPFS for decentralized content storage creates a powerful and efficient system for managing verifiable information.

Figure 5.2 shows how the framework integrates DIDs, smart contracts, and IPFS to enable a decentralized trust layer for content provenance and verification.

## 5.4 FRAMEWORK WORKFLOW AND AI INTEGRATION

Following the description of the main components of the framework, this section describes its operational workflow. This paper explains the full lifecycle for creating content, authentication, and dissemination in this decentralized ecosystem, with a specific focus on how the framework counteracts GenAI-specific misinformation. Furthermore, it describes the strategic integration of AI as an auxiliary layer to enhance the framework's capabilities without compromising its decentralized features.

### 5.4.1 Content lifecycle: creation, verification, and dissemination workflow

The effectiveness of the framework depends on the life cycle of the precisely designed content that ensures the origin and integrity from the start to consumption. This workflow is structured to integrate decentralized technologies in critical situations, reinforcing trust in each phase.

As shown in Figure 5.3, the content-validation framework converts content into a decentralized and multistage process that ensures unity of the creation and consumption.

The process begins with the creation of primary content. When a content maker (e.g., a journalist, a researcher, or an individual citizen) generates a digital content, they begin their entry into an acceptable ecosystem. Utilizing their DID, the creator cryptographically signs the content. A unique cryptographic hash is then generated from the content and immediately anchored

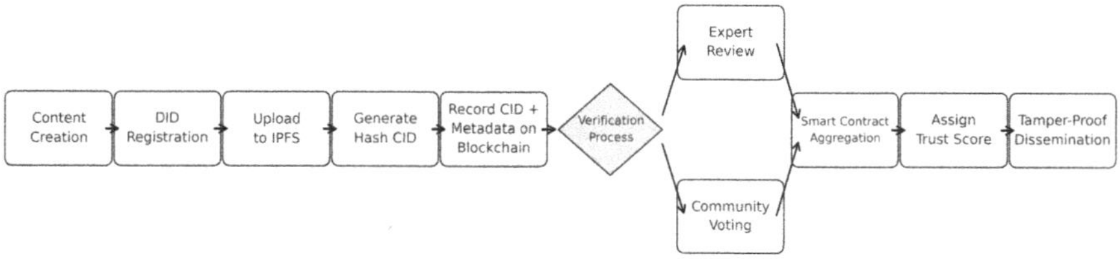


*Figure 5.3* The decentralized verification workflow.

to the underlying blockchain via a smart contract. The actual content file is stored on the IPFS, with only its hash and the creator's DID being immutably recorded on the ledger. This initial reliance thereby establishes the origin of the acceptance and authorship proof. For instance, a news agency publishing an article would hash the article and its associated media, anchoring these hashes to the blockchain, thereby creating an immutable record of its initial publication (England et al., 2020).

After creation, the content enters the decentralized approval phase. Unlike traditional systems that rely on a centralized system, this framework employs a distributed network of verifiers. These verifiers, which have DIDs, can be accessed by content (via hash functions) and its blockchain-anchored metadata. Confirmation can include different methods, such as cross-referencing with other verifiable sources, using specialized analytical tools, or involving community consensus mechanisms. Smart contracts facilitate the process by managing authentication tasks, recording authentication results, and updating the trust score or the status. For example, multiple independent verifiers may examine a controversial part of the content, and their collective assessments, signed cryptographically, may be recorded on blockchain and may affect the extent of public trust in the content.

The final stage of emission and consumption is safe. When the content is confirmed and its integrity established (or flagged if found to be misleading), it can be circulated securely among different platforms. Programs and user interfaces can be integrated with stakeholders to retrieve their content and validate its authenticity before the general ledger is changed before presentation to users. This allows consumers to not only see content but also see the full endorsement of it, including authorship, breeding charts, and results. This model enables users to make informed decisions about the information they use and provides an increasing number of informative readers with correct information.

### 5.4.2 Mechanisms for combating GenAI-specific misinformation

This framework is specifically designed to address the unique challenges arising from inaccurate information generated by GenAI and moving beyond public content verification for targeted cross-purpose actions. The complexity of artificial media requires expert identification and detection of the mechanisms.

One of the main mechanisms of detection is artificial media. By applying the original content at its creation point, the framework seeks to create a continuum for all digital media. For instance, cameras or recording devices could be equipped with hardware-level cryptographic modules that automatically hash and sign media files as they are captured, embedding this immutable fingerprint onto the blockchain. This "resource confirmation" makes it significantly more difficult to enter the ecosystem because each media without the origin of authentication or lack of coordination in its arrest chain is immediately labeled suspicious (Calzada et al., 2025). This preventive approach shifts the load from reactive recognition to authenticity from the beginning.

Table 5.1 shows how the proposed framework enables traceability and authentication for content generated by both AI and by human intelligence, using anchors and content hashes.

Another critical mechanism involves AI-Generated Textual Content Detection and Attribution. While AI can generate highly convincing text, subtle linguistic patterns or metadata anomalies can often betray its synthetic origin. The framework can integrate specialized AI classifiers (discussed further in Section 5.4.3) designed to identify these patterns. More importantly, by linking content to DIDs, even if the text is AI-generated, the framework allows for the attribution of the content to the entity that deployed or prompted the AI model. This provides an important layer of accountability to ensure that false information generated by AI cannot hide behind the anonymity of machine writing. The focus shifts from merely identifying "if it's AI-generated" to "who is responsible for this AI-generated content."

Moreover, the frame supports the detection of automatic and decreasing anomalies. Smart contracts can be programmed to monitor suspicious activities, such as quickly posted content from new platforms, content with low repute, or sudden changes in content without corresponding corrections. These automated warnings can increase the accuracy of human or AI tools, providing an initial warning system against a consistent lie information campaign. This capability of preventive control significantly reduces the time window for the diffusion of incorrect information before the intervention.

*Table 5.1* Provenance comparison: AI-generated vs. authentic content

| *Feature* | *AI-generated content* | *Authentic content* |
|---|---|---|
| Provenance | Model ID+User DID signed | Creator DID+Timestamp |
| Modification | Hash mismatch triggers alert | Original hash remains constant |
| Detection | AI analysis+ZKP (optional) | Direct blockchain verification |

### 5.4.3 Leveraging AI as an assistive layer in verification

While the framework's core trust mechanism is rooted in blockchain's decentralization and immutability, AI plays a crucial, yet carefully defined, assistive role. AI is not used as a focused referee of truth but as a powerful tool to enhance human capabilities and increase the efficiency of decentralized authentication processes. This hybrid approach is based on the analytical capabilities of AI and reducing the risk of algorithmic bias and centralized control.

One of the key applications is the analysis and processing of AI-based pre-screening. Before the content enters the decentralized authorization workflow, AI models can perform an initial scan for typical indicators of incorrect information, such as light anomalies, real contradictions, or artificially generated signatures. For example, natural language processing (NLP) models can analyze textual content for patterns that represent AI or deceptive language, while computer vision algorithms can analyze images and videos of artifacts and deepfake (Paul, 2024). This processing helps prioritize the content for human review, reduce workload, and accelerate the overall process. It acts as a smart filter, not the final judge.

One of the crucial roles of AI is recognizing reputation and recognizing anomalies for verifiers. AI algorithms can analyze the historical performance of verifiers in the network, evaluate the accuracy of past judgments, and assess their consistency with the society's consensus. This can contribute to a dynamic reputation system where accurate and reliable verifiers have more influence in the certification process. Similarly, AI can recognize unusual patterns in authentication behavior such as a coordinated reduction or attempt to manipulate the scores of trust and thus maintain the integrity of the decentralized confirmation community. This application increases the nature of the decentralized network.

Finally, AI can be used for information synthesis. In a complex information environment, providing a comprehensive database of users around a piece of content is very important. AI can quickly provide information from multiple reliable sources, summarize different views, or highlight relevant background details that help users assess content effectively. This is beyond simple fact control—providing a richer understanding of the information landscape enables users to form their conscious thoughts. This solution ensures that AI supports human decision-making rather than replacing it.

## 5.5 PRACTICAL APPLICATIONS AND ETHICAL CONSIDERATIONS

According to the theoretical framework and operational design, this section reviews practical applications. We are talking about these blockchain-based trust settings and how we can control information and data, especially when AI pulls out incorrect information everywhere. At the same time, we deal with big moral questions and the massive waves of society that will be dealt with by releasing what the game changes. Everything about the emphasis on intelligent monitoring and the maintenance of technological advances is under balanced control.

### 5.5.1 Practical use cases: news verification, deepfake provenance, and beyond

The framework will have a serious impact on real-world scenarios and change the script about how we confirm and trust in different fields. Its main strengths—like rock-solid origin tracking, spread-out checks, and crypto-linked ownership—tackle some of the biggest issues in our digital info world head-on.

As of now, one of the biggest wins is in News Verification and Journalistic Integrity. With everyone second-guessing what's real in the news these days, this system lets reporters and outlets lock their articles, reports, and media onto the blockchain the second they're created, which undoubtedly eliminates a major issue. Any changes will come into the system later, and everyone can audit. Reading people can use an easy means to check the background history and make sure that nothing has happened since the beginning. This will not only restore faith in the true news, but it will also give journalists a shield to protect their work from false improvements or credibility issues.

Another important application is in the origin of artificial media. As mentioned in the previous chapters, the ability of GenAI to create deepfakes poses a severe threat. The framework acts as a direct counter to the creation of an authentication chain for digital media from its source. Cameras or recording devices place cryptographic hashes directly on a blockchain at the moment of creation. This "birth certificate" enables the media to detect valid content from bogus copies. Each deepfake lacks such a reliable source or someone whose identity does not match the blockchain record can be immediately flagged as suspicious. This shifts the paradigm from trying to identify after its spread to actively confirming the authenticity of all media from its origin.

Beyond these critical areas, these ideas that spread to different areas consider academic credit and the integrity of the research: the framework can lock in invariant precedents for degrees, chapters, and data; stop fraud; and ensure realistic roots of work (Alshahrani et al., 2022) or transparency of the supply chain: tracking the goods from the beginning to the shelf is morally acceptable, though that's more of a general blockchain perk beyond just fake news (Queiroz et al., 2019). Even in Digital Art and NFTs, it can protect artists and shut down fake works. All of these points show how flexible and robust a decentralized trust system can be.

### 5.5.2 Balancing decentralization with responsible governance

The power of decentralization, while offering many benefits for resilience, poses complex challenges, especially in relation to governance and accountability. An effective framework should guide the subtle balance between distributed control and mechanisms that ensure responsible action and resolution.

The inherent nature of decentralized systems means there is no single central authority to enforce rules or mediate conflicts. This necessitates the implementation of decentralized governance models, often leveraging smart contracts and community consensus mechanisms.

As depicted in Figure 5.4, the framework must navigate inherent tensions between competing ethical priorities, which requires careful and context-sensitive balancing.

Concepts such as Decentralized Autonomous Organizations (DAOs) can play a pivotal role, allowing the network participants to vote on protocol upgrades, content reduction policies, and dispute resolution methods (Davidson et al., 2016). The purpose of these models is to distribute decision-making power and increase the system's resistance to capture or penetration. However, the design of decentralized and fair governance is a complex act that requires careful consideration of voting mechanisms, incentive structures, and representation to prevent the emergence of new forms of focus.

Furthermore, the framework shall address the challenge of accountability in a pseudonymous environment. While DIDs provide authentication, many blockchain systems allow identity creation, meaning users can function

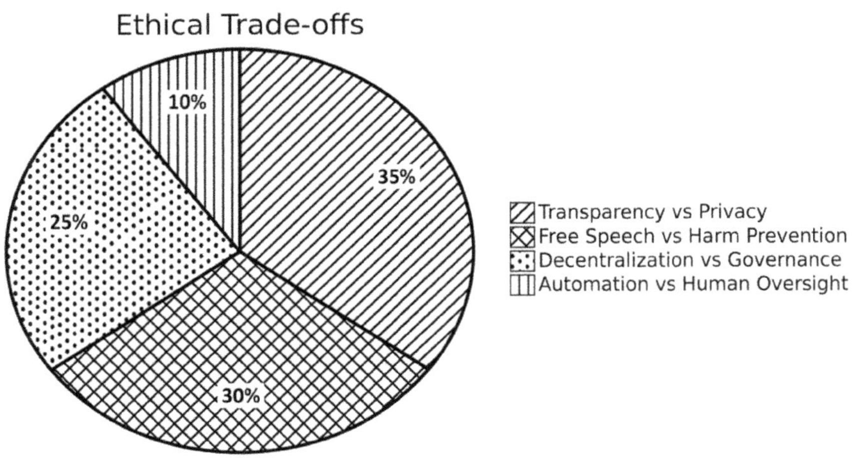


*Figure 5.4* Key ethical trade-offs in decentralized verification systems.

without revealing their true identity. This can be useful for privacy, but it complicates accountability when destructive operators use the system. This framework should carry out strong credit systems in which DIDs gather privileges of trust based on their historical assistance to authenticate and integrate content. These reputation scores, managed by smart contracts, can influence the voting power of participants or access to specific functions and provide a decentralized mechanism to respond without compromising privacy. This balance is necessary to foster a self-contained society.

### 5.5.3 Ethical implications and societal impact

The implementation of the blockchain framework for information containing ethical concepts and significant social impacts on the society requires that we face it before we do harm. Of course, this technology is built for crushing erroneous information, but if we are not careful, it can create new problems or create wide gaps in society.

The main concern is privacy and transparency. The system seeks transparency through immutable content provenance, but this must be balanced with individuals' right to privacy. For example, while tracking the resources of content is critical, we cannot allow personal details or sensitive metadata to disappear, which requires strict privacy tools such as zero-knowledge proofs or selective disclosure of information. These settings should allow users to decide what is public and to abide by rules such as the use of "required" and "satisfied."

Another important aspect is the potential for misuse and abuse. Even with decentralization, it could get twisted. For example, large players may be "washing the truth" with the rigging of a class of hacked DIDs to verify or close the true information with false flags of the gang. This framework needs strong anti-team defense systems and flexible agent systems to identify and prevent these threats. At the top, the idea of the "verifiable fact" can be turned into weapons if authority disappears, triggers censorship, or undermines different views (O'Neil, 2016).

To remove these risks, the defense framework layers against various biases are shown in Table 5.2.

The social impact on trust and the digital divide also requires attention. While the purpose of this framework is to restore trust, its

*Table 5.2* Types of bias and mitigation strategies

| *Bias type* | *Source* | *Framework mitigation strategy* |
|---|---|---|
| Algorithmic | Training data imbalances | Explainable AI (XAI) audits |
| Human moderator | Cultural/political leanings | Diverse verification pools with blind voting |
| Network effects | Dominance of early adopters | Reputation decay mechanisms |

acceptance and effectiveness depend on the availability of digital literacy. If access to the framework is limited by technological barriers or economic disparities, it can inadvertently aggravate the digital gap and create a two-storey information ecosystem that only privileged users can access the reality. Efforts should be made to provide support for the framework that is user-friendly, adaptable to different platforms, and accessible to the global audience, in order to promote fair access to reliable information. Finally, the success of this framework will be measured not only by its technical prowess but also by its ability to foster a conscious, flexible, and fair society.

## 5.6 CHALLENGES, FUTURE DIRECTIONS, AND CONCLUSION

The early sections of this chapter combine a set of solid blockchain configurations to make digital information more reliable in this GenAI world. The initial concepts and design seem like winners, but commencing it and integrating into everyday use might hit some real snags. In conclusion, I'll break down those major roadblocks, point out some cool avenues for where things could head next in terms of research and tweaks, and tie it all up with a quick recall at what this framework really offers and the bigger picture it paints for a tougher, more honest online space.

Figure 5.5 shows the evolution of the stage needed to address key challenges that provide a roadmap for the maturity of the framework.

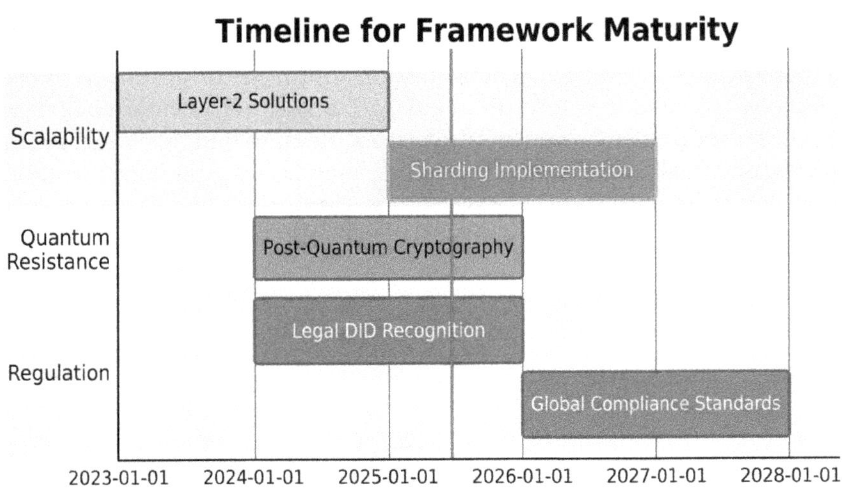


*Figure 5.5* Roadmap for framework maturity and overcoming key challenges.

## 5.6.1 Scalability, interoperability, and adoption challenges

It's not easy to set up a wide-ranging system to check information around the world—it's full of hard spots, especially in technology, such as large-scale management, linking different platforms, and convincing people to jump. Nailing them is what makes it a smart app that changes the game.

Scale is the most stubborn issue. We rely on the IPFS to store large files, but the flood of hashes and verification steps can still overwhelm things. Options such as layer 2 (e.g., lightning or polygon) or division of work volume (Ethereum 2.0 style) seem promising, but they are not yet fully present and require more polishing and real-world experiments. The digital scenario is a chaos of platforms—social networks, news centers, and content tools—you name it. For this blockchain approach to have remarkable effects, it must be mingled with all the chaos. This means enabling universal APIs, common methods for data control, and bridges between chains or even old systems for information exchange. Without this interoperability, the framework risks becoming an isolated island disconnected from the broader digital ecosystem, severely limiting its potential reach and adoption.

As Table 5.3 shows, the emerging standards and protocols are providing pathways to enable this vital collaboration capability.

User acceptance and usability are perhaps the most important nontechnical challenges. The best technology in the world is not important if it is confusing or not worth the effort to create the same browser and browsers. Jargon such as blockchain, crypto hoops, and identifier management can scare the average person. We need simple schemes to resolve the complexity and let people focus on what is important and help people appreciate and value real things, and we have to change those deeply rooted habits that pass through the junk. The creation of faith in this decentralized change requires intelligent training and a smooth ride from start to finish.

Finally, we cannot ignore the regulatory and governmental challenges. This decentralized environment often treats the rules built for large, central players, and questions about data control, lying blame, and implementation throughout the world are intertwined in a web like this. Creating clear,

*Table 5.3* Examples of cross-chain solutions for interoperability

| *Protocol* | *Approach* | *Throughput* | *Use case* |
|---|---|---|---|
| Polkadot | Parachains + relay chain | 1,000–10,000 TPS | Multi-chain verification systems |
| Cosmos | IBC (inter-blockchain comm.) | ~1,000 TPS | Enterprise blockchain networks |
| Chainlink CCIP | Oracle-based bridges | Variable | DeFi and data provenance |

accepted means for groups such as DAOs, while avoiding a global patchwork, will be very difficult to keep it above board and make public.

### 5.6.2 Future synergies: explainable AI (XAI) and quantum-resistant blockchains

This trust framework on blockchain is bound to evolve hand-in-hand with breakthroughs in AI and quantum stuff. Digging into how these can team up could increase its strength and contribute to staying power in big ways.

Explainable AI (XAI) is one exciting path to blend AI deeper but with way more clarity into the checking process. Right now, AI pitches in as a helper, but its discreet inner mechanisms can leave folks doubting, especially on high-stakes calls like spotting fakes. XAI flips that by laying out the logic behind decisions—say, spelling out why a video got tagged as manipulated. Weaving this into our AI tools would boost confidence in the outputs and give human reviewers the scoop to grasp and maybe adjust those calls. It's all about that transparency, which lines up perfectly with the framework's push for things you can double-check and rely on.

Another critical future consideration is the development of Quantum-Resistant Blockchains. The security of current blockchain cryptography, particularly public-key encryption schemes, relies on the computational difficulty of certain mathematical problems. However, the advent of powerful quantum computers in the future could potentially break these cryptographic foundations, rendering existing blockchains vulnerable to attacks. Research into "post-quantum cryptography" is actively developing new cryptographic algorithms that are resistant to quantum attacks. For the long-term resilience and future-proofing of the decentralized trust framework, transitioning to or integrating with quantum-resistant blockchain protocols will be essential to ensure the continued immutability and security of content provenance records against future threats.

### 5.6.3 Concluding remarks

With GenAI providing misinformation like there's no tomorrow, declining trust in what we see online poses a serious risk to how democracy is managed and preserved. This chapter sketches out a blockchain-powered way to decentralize what's true and inject some real integrity back into our digital lives. Due to intensive tracking, extensive identification cards, and intelligent transactions, in addition to IPFS for smart storage, this system challenges the opaque and manipulative systems that currently enable the proliferation of fake content.

As discussed in Table 5.4, the synergistic architecture of the framework also addresses both immediate and emerging threats to information integration.

*Table 5.4* Quantified benefits of the proposed framework

| *Feature* | *Benefit* | *Impact* |
|---|---|---|
| Decentralized identifiers (DIDs) | Eliminates impersonation and enhances source authenticity | Estimated 90% reduction in fake accounts |
| AI-blockchain integration | Combines analytical capabilities with tamper-proof auditability | 3× improvement in deepfake detection speed |
| Quantum-resistant cryptography | Ensures future-proof security against quantum threats | Sustains long-term cryptographic integrity |

We have designed a complete content lifecycle — from initial hashing and anchoring, to decentralized source verification, and the use of secure, privacy-preserving sharing tools — to combat the spread of deepfakes and AI-generated misinformation. AI is introduced as an amplifier, not the primary decision-maker, enhancing speed through automated screening, trust scoring, and contextual analysis, while keeping humans in control and distributing decision-making across the network. Certainly, barriers such as growing problems, communicating with users, and their linkage, in addition to persistent ethical dilemmas and malicious actors, remain significant hurdles. However, planned integrations with Explainable AI (XAI) and quantum-resistant cryptography offer promising pathways to strengthen and future-proof the framework.

At its core, this framework is about empowering everyday people with verifiable facts, dispelling confusion in our online world, and uniting all stakeholders to resist the spread of false information. Achieving this will require collaboration from everyone — technologists, policymakers, content creators, and users from all walks of life — to get on board with a fresh take where trust isn't handed down from on high, but built from the ground up through checkable trails and shared agreement.

## REFERENCES

Aguerre, C., Campbell-Verduyn, M., & Scholte, J. A. (2024). *Global Digital Data Governance: Polycentric Perspectives*. Routledge.

Allen, D., & Weyl, E. G. (2024). The real dangers of generative AI. *Journal of Democracy, 35*(1), 147–162.

Alshahrani, M., Beloff, N., & White, M. (2022).Towards a blockchain-based smart certification system for higher education: An empirical study. *International Journal of Computing and Digital Systems*, 11(1), 553–571. https://doi.org/10.12785/ijcds/110145

Alwaisi, S., Al-Radhi, M. S., & Nemeth, G. (2023). Automated child voice generation: Methodology and implementation. In *Proceedings of the 2023 International Conference on Speech Technology and Human-Computer Dialogue (SpeD)*, Bucharest, Romania (pp. 48–53). IEEE.

Amoore, L., Campolo, A., Jacobsen, B., & Rella, L. (2024). A world model: On the political logics of generative AI. *Political Geography, 113*, 103134.
Angelidou, M., & Sofianos, S. (2024). The future of AI in optimizing urban planning: An in-depth overview. In *Book of Abstracts, Changing Cities VI Conference*, Rhodes, Greece.
Calzada, I. (2021). *Smart City Citizenship*. Elsevier.
Calzada, I., Németh, G., & Al-Radhi, M. S. (2025). Trustworthy AI for whom? GenAI detection techniques of trust through decentralized Web3 ecosystems. *Big Data and Cognitive Computing, 9*(3), 62. https://doi.org/10.3390/bdcc9030062
Chafetz, H., Saxena, S., & Verhulst, S. G. (2024). A fourth wave of open data? Exploring the spectrum of scenarios for open data and generative AI. The GovLab. arXiv:2405.04333
Cugurullo, F., et al. (Eds.). (2024). *Artificial Intelligence and the City: Urbanistic Perspectives on AI*. Routledge.
Davidson, S., De Filippi, P., & Potts, J. (2016). Economics of blockchain and DAOs. *SSRN Electronic Journal*. https://doi.org/10.2139/ssrn.2744751
Delacroix, S. (2024). Sustainable data rivers? Rebalancing the data ecosystem that underlies generative AI. *Critical AI, 2*(1). https://doi.org/10.1215/2834703X-11205224
England, P., et al. (2020). AMP: Authentication of media via provenance. arXiv preprint arXiv:2001.07886v3
Europol. (2022). *Facing Reality? Law Enforcement and the Challenge of Deepfakes*. Publications Office of the European Union.
European Commission. (2024). The future of European competitiveness: A competitiveness strategy for Europe. Retrieved November 18, 2024, from https://commission.europa.eu/topics/eu-competitiveness/draghi-report_en
European Parliament and Council. (2024). Regulation (EU) 2024/1689 of 13 June 2024 laying down harmonised rules on artificial intelligence. *Official Journal of the European Union, L1689*, 1–144.
Fang, R., et al. (2024). LLM Agents can autonomously hack websites. arXiv preprint arXiv:2402.06664
Farina, M., Yu, X., & Lavazza, A. (2023). Ethical considerations and policy interventions concerning the impact of generative AI. *AI & Ethics, 5*, 737–745.
Gabriel, I., et al. (2024). The ethics of advanced AI assistants. arXiv preprint arXiv:2404.16244
Insight & Foresight. (2024). How generative AI will transform strategic foresight. Retrieved from https://hkifoa.com/wp-content/uploads/2024/12/how-genai-transform-strategic-foresight.pdf
Kitchin, R. (2014). *The Data Revolution: Big Data, Open Data, Data Infrastructures and Their Consequences*. SAGE Publications.
Liang, W. (2025). Smart Contracts and Corporate Governance: Automation, Legal Risks, and Benefits.
Mafrur, R. (2025). AI-based crypto tokens: The illusion of decentralized AI? arXiv preprint arXiv:2505.07828v1
O'Neil, C. (2016). *Weapons of Math Destruction: How Big Data Increases Inequality and Threatens Democracy*. Crown Publishers.
Park, P. S., et al. (2025). AI deception: A survey of examples, risks, and potential solutions. *Patterns, 5*(5), 100988. https://doi.org/10.1016/j.patter.2024.100988

Paul, J. (2024). The role of AI in combatting misinformation and fake content [ResearchGate preprint]. Retrieved from https://www.researchgate.net/publication/387086008

Politou, E., Casino, F., Alepis, E., & Patsakis, C. (2019). Blockchain mutability: Challenges and proposed solutions. *IEEE Transactions on Emerging Topics in Computing*, 9(4), 1972–1986. https://doi.org/10.1109/TETC.2019.2949510

Queiroz, M. M., Telles, R., & Bonilla, S. H. (2019). Blockchain and supply chain management integration: A systematic review of the literature. Supply Chain Management: *An International Journal*, *25*(2), 241–254. https://doi.org/10.1108/SCM-03-2018-0143

Saif, M. B., Migliorini, S., & Spoto, F. (2024). Efficient and secure distributed data storage using IPFS and blockchain. *Future Internet, 16*(2), 98.

Shin, D., Koerber, A., & Lim, J. S. (2024). Impact of misinformation from generative AI on user information processing. *New Media & Society*, *27*(7), 4017–4047.

Tsai, L. L., et al. (2024). *An MIT Exploration of Generative AI: From Novel Chemicals to Opera.* MIT Governance Lab.

Voshmgir, S. (2020). *Token Economy: How the Web3 Reinvents the Internet* (2nd ed.). Token Kitchen.

Weidinger, L., et al. (2023). Sociotechnical safety evaluation of generative AI systems. arXiv preprint arXiv:2310.11986

Wu, B., & Seneviratne, O. (2025). Blockchain-based framework for scalable and incentivized federated learning. arXiv preprint arXiv:2502.14170v1

Yang, F., Goldenfein, J., & Nickels, K. (2024). *GenAI Concepts.* RMIT University, ARC Centre of Excellence for Automated Decision-Making and Society.

Chapter 6

# An AI-augmented human-in-the-loop (HITL) approach for misinformation prevention in the digital age

*Pijush Dutta, Rahul Vadisetty, Amrita Mitra, Kakali Das, Atri Adhikari, and Jyoti Sekhar Banerjee*

## 6.1 INTRODUCTION

To improve false information detection and mitigation, the artificial intelligence (AI)-augmented Human-in-the-Loop (HITL) platform of misinformation prevention combines human knowledge and machine learning (ML) (Aïmeur et al., 2023). The merits of human judgment and AI capabilities are used in this hybrid model, which allows addressing the problems of the scalability characteristic of traditional fact-checking. Key elements of this approach are influenced by human expertise integration and the Feedback Mechanism (Ardia et al., 2020). The SAMS-HITL model enables users to give feedback on misinformation, which makes them better self-fact-checkers, which increases the overall accuracy by 7.1% (Shabani et al., 2021).

Crowdsourcing: Experts' input, intertwined with the crowdsourcing data, will result in a more powerful system to check the facts and provide real-time evaluations of the information's validity (La Barbera et al., 2022). AI and Technologies: Real-Time Detection: AI systems are able to cross-verify news sources and patterns in language, and with an average accuracy of up to 89%, fake news can be detected within a low processing latency (Chen et al., 2024). The hybrid model Bidirectional Encoder Representations from Transformers-Long Short-Term Memory (BERT-LSTM) has a high accuracy (93.51) in the detection of misinformation, and it is also a teaching tool that can be used to improve digital literacy (Dong et al., 2024).

Natural language processing (NLP), multimodal content categorization, and machine text analysis have all seen remarkable advancements in AI (Sriharsha et al., 2021). Yet, AI-only systems with a full set of automation are highly problematic, where the AI-based models are commonly confused by sarcasm, cultural backgrounds, or hidden misinformation. The creators of misinformation keep evolving their strategies; therefore, it is hard to keep up with the unchanging AI models (Alam et al., 2025).

Problems of bias and fairness: AI models are trained on biased data, and consequently, they are likely to promote unfair or wrongful classifications. Automated systems can be characterized as black boxes, limiting the trust they generate (Li, 2024; Rao et al., 2025). These shortcomings indicate that AI

DOI: 10.1201/9781003626558-8

alone cannot ensure the accuracy and reliability required for robust misinformation prevention. HITL integrates human judgment with AI efficiency, creating a hybrid decision-making system. Humans provide critical insights that AI may overlook, such as cultural context, ethical reasoning, and ambiguous cases where automated detection is uncertain (Kumar et al., 2024). HITL ensures higher accuracy by validating AI predictions, facilitating continuous learning as human feedback improves AI models, promoting trustworthiness through human oversight that adds transparency to automated decisions, and enhancing adaptability as humans can identify novel misinformation patterns faster than AI retraining cycles (Emami et al., 2024; Enarsson et al., 2022). Thus, HITL plays a vital role in achieving a balanced, reliable, and ethical approach to misinformation prevention.

The motivation for this research arises from the urgent need to counter misinformation effectively in digital ecosystems. Purely AI-driven systems lack reliability, while human-only moderation is time-consuming, costly, and unscalable. This study proposes an AI-augmented HITL framework that combines the speed and scalability of AI with the judgment and contextual awareness of humans.

The main objectives are as follows:

- To design a hybrid misinformation detection framework that integrates AI detection and human validation.
- To evaluate the effectiveness of HITL in improving accuracy and reducing false positives/negatives.
- To demonstrate how continuous human feedback can retrain and enhance AI models over time.
- To explore the practical applications of such systems in social media platforms, news verification, and digital communication ecosystems.

This paper makes the following contributions:

It proposes a novel AI-augmented HITL architecture for misinformation prevention in the digital age and outlines a stepwise process including input collection, preprocessing, AI detection, HITL verification, decision-making, and feedback-driven retraining. Reliability enhancement demonstrates how HITL addresses limitations of AI-only systems, particularly in handling ambiguity, bias, and explainability issues (Moreno-Sánchez et al., 2025) and provides a foundation for scalable misinformation prevention systems that evolve with new forms of digital misinformation.

Practical Implications: It highlights the applicability of the proposed framework for governments, digital platforms, and researchers working on misinformation management.

## 6.2 RELATED WORK

AI has been widely employed to detect misinformation using NLP, ML, and Deep Learning (DL). Text-based approaches rely on linguistic features, sentiment analysis, stylistic cues, and semantic similarity to identify misleading information (Poojary, 2024). Transformer-based models such as BERT, RoBERTa, and GPT are increasingly used for fake news detection due to their ability to capture contextual meaning. Knowledge graphs, databases, and external fact-check repositories are used by fact-checking systems to verify claims based on trusted sources. Multimodal approaches integrate text, images, and videos in detecting fake pictures, edited videos, and deepfakes. Regardless of these improvements, AI models are frequently

*Table 6.1* State-of-the-art techniques for fake news detection

| *Reference* | *Summary* | *Outcome* |
|---|---|---|
| Shabani et al. (2021) | SAMS-HITL is a machine learning and human fact-checking algorithmic approach that is used to improve the detection of misinformation. It uses human feedback about the source, author, message, and spelling, which makes it more accurate in classification (7.1% compared to automated methods alone) | The SAMS-HITL approach improves accuracy in classifying news articles as true or false by up to 7.1%. |
| Caldarelli (2025) | The article is concerned with a real-time intelligent model of detecting and mitigating fake news by a multisource verification instead of a Human-in-the-Loop (HITL) solution. It focuses on the automated cross-verification and the use of language pattern analysis to increase the accuracy of information efficiently. | Achieved 89% accuracy in detecting fake news and processing latency lower than 1.5 seconds per article. |
| Sriharsha et al. (2021) | The article addresses a BERT-LSTM-based fake news detection model, with a high accuracy and ability to create digital literacy, yet it lacks a HITL framework. | The detection of misinformation in the proposed model was 93.51% accurate, 91.96% recall, and 92.73% F1. The respondents of the experimental group recorded higher confidence levels and a shorter time of decision-making. |
| Rahaman et al. (2024) | The paper highlights the necessity of cooperation among technologists, researchers, and policymakers, whereby an AI-Augmented HITL methodology should be adopted to better prevent misinformation so that the potential of AI could be fully harnessed in addition to human judgment to create a more resilient online environment. | AI serves as both creator and detector of deepfakes. Collaboration enhances credibility and combats digital misinformation effectively. |

*(Continued)*

*Table 6.1* (*Continued*) State-of-the-art techniques for fake news detection

| *Reference* | *Summary* | *Outcome* |
|---|---|---|
| La Barbera et al. (2022) | Combining AI methods, crowdsourcing, and professional involvement, the article proposes a hybrid human-in-the-loop system for fact-checking that effectively counteracts online disinformation and becomes more scalable and effective in validating the correctness of information in the digital era. | The paper proposes an alternative human-in-the-loop fact-checking system that uses crowdsourcing, autonomous AI methods, and human involvement. It addresses the issue of disinformation through analyzing the interaction of the different components of the framework. |
| Traberg et al. (2023) | The article is silent on an AI-enhanced HITL strategy in the prevention of misinformation. It concentrates on psychological inoculation interventions and digital age gamified measures to fight misinformation, specifically in the context of COVID-19 vaccine hesitancy. The paper gives an overview of a long-term solution to the problem of countering misinformation. The paper has brought to attention the scientific developments in using psychological inoculation interventions. | The paper presents the upcoming trends in the fight against fake news, such as the analysis of user behavior and joint cybersecurity measures, which are consistent with AI Augmented HITL. The approaches improve the performance in detection and minimize limitations such as adversarial attacks and biased training samples. Weaknesses and strengths of AI detection approaches. Figure out recognized new trends in fake news detection tactics. |
| Vishnupriya et al. (2024) | In the paper, the author speaks about the insufficiency of conventional ways of misinformation detection in the age of generative AI models and why a systematic, explainable solution is required, including both psychological and societal insights, which can be effectively used to tackle AI-generated misinformation. | Evaluated strengths and limitations of AI-based detection methods. Identified emerging trends in fake news detection strategies. |
| Xu et al. (2023) | The paper discusses the inadequacy of traditional misinformation detection methods in the era of generative AI models, emphasizing the need for systematic, explainable solutions that incorporate psychological and societal perspectives to effectively combat AI-generated misinformation. | Identifying manipulation traces in AI-generated content. Providing a systematic, explainable solution to combat misinformation. |

susceptible to misinformation that is context-specific, e.g., satire, sarcasm, and location-specific cultural references. HITL techniques have become a necessary addition to AI-based detection. In content moderation, Facebook, Twitter (X), and YouTube use human reviewers to approve or abide by the content in question, which has been flagged by AI. Likewise, to verify news articles and viral claims presented in Table 6.1, fact-checking organizations (e.g., PolitiFact, Snopes, and FactCheck.org) also use the help of human experts.

HITL has a number of advantages, such as offering contextual meaning that AI is incapable of. It guarantees responsibility and ethical control over content moderation and allows end-to-end human-verified feedback loops to continuously improve AI models. Nevertheless, HITL also presents issues such as the cost, scalability, or possible fatigue or bias of the reviewer. The AI-only scalable automated systems are effective in terms of speed, scalability, and real-time detection but have flaws of false positives, a lack of contextual judgment, and evolving misinformation tactics. Although human moderation can be accurate, it is sluggish, labor-intensive, and ineffective in large-scale digital ecosystems. Hybrid AI+HITL methods represent a balance between the efficiency of AI and human judgment since they can be scalable and reliable at the same time. On one hand, such systems are more efficient in unclear or risky situations, in which AI is not sure. Research shows that the AI+HITL models perform better than single AI systems in terms of accuracy and reliability and thus are a good fit in misinformation prevention in dynamic online settings.

## 6.3 THEORETICAL FOUNDATION

It is also important to be aware of the different types of misleading information to develop effective prevention models. In misinformation, there is the sharing of false or inaccurate information without any ill intent. For example, a user may unknowingly forward outdated health advice (Bhatta, 2024).

Disinformation: Deliberately false or manipulated content shared with the intention to deceive, mislead, or cause harm. This includes political propaganda, fake news campaigns, or deepfake videos. Malinformation: Genuine information used in a misleading or harmful context, such as selectively publishing private data or presenting facts in a distorted manner. These distinctions are critical because each type requires a different detection and intervention strategy, highlighting the need for adaptive and context-aware frameworks (Collins et al., 2021; Hangloo & Arora, 2022).

AI provides the first layer of defense by automatically identifying suspicious content through various techniques. In NLP, it detects linguistic cues such as exaggeration, bias, or inconsistencies, uses transformer-based models (BERT, RoBERTa, and GPT-like systems) for semantic analysis, and identifies sentiment polarity and contextual meaning in news articles, posts, and comments (Gautam et al., 2025). Multimodal learning analyzes text, images, and videos together to identify manipulations. Deep learning models convolutional neural networks (CNNs), recurrent neural networks (RNNs), and Vision Transformers) detect deepfakes, doctored images, and synthetic media (Rajput & Arora, 2024). Fact-checking algorithms, compare claims against knowledge graphs, trusted fact-check databases, and external references and use semantic similarity and entity recognition to determine claim validity. Together, these AI techniques act as automated gatekeepers, but they often produce errors in ambiguous cases or evolving misinformation scenarios, and the HITL framework paradigm integrates human expertise into the AI workflow to overcome the limitations of fully automated systems (Kumar et al., 2024). In this framework, AI performs initial filtering and risk scoring of suspicious content. Human reviewers evaluate flagged cases, applying contextual, cultural, and ethical reasoning. Human feedback is used to validate AI predictions and reduce false positives/negatives. This creates a feedback loop where human judgments continuously improve AI performance through retraining. The HITL framework ensures greater accuracy, accountability, and trustworthiness in misinformation detection, especially in high-risk or ambiguous scenarios (da Silva, 2024). Justification for Combining AI with Human Expertise: The combination of AI and human intelligence is not only beneficial but also necessary for scalability, speed, and the ability to process massive volumes of data in real-time, as well as cultural sensitivity, ethical discernment, contextual interpretation, and alignment with emerging disinformation trends.

It reduces massive data streams, and human beings are used to give ultimate validation and to make decisions with subtlety. This compromise stops excessive censorship and misclassification, as well as the propagation of false information. Further, the integration of AI and human knowledge can help overcome ethical issues, increase explainability, and make the system transparent and responsible. Thus, the theoretical foundation for this work is rooted in the complementary strengths of AI and human judgment, forming a resilient and adaptive misinformation prevention strategy.

## 6.4 PROPOSED METHODOLOGY: AI-AUGMENTED HITL FRAMEWORK

The proposed framework follows a hybrid AI-HITL pipeline shown in Figure 6.1. At a high level, it consists of the following modules:

### 6.4.1 Input and data collection

The system ingests digital content from multiple platforms, including the following: Social media posts (Twitter/X, Facebook, and Instagram), news articles from verified/unverified outlets, user-generated content (comments,

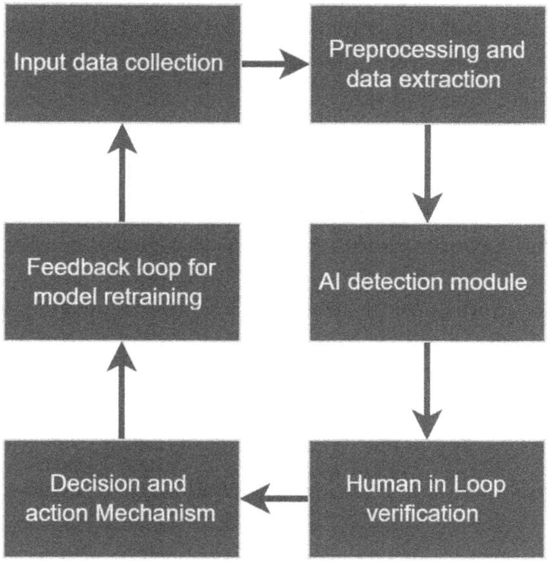


*Figure 6.1* Proposed framework, hybrid AI-HITL.

blogs, and videos), multimedia formats (text, images, audio, and video), and application programming interfaces (APIs), scrapers, or platform-level integration may be used to collect real-time or batch data for analysis.

### 6.4.2 Preprocessing and feature extraction

Before analysis, raw data are standardized and structured by using text preprocessing that is done by tokenization, stopword removal, stemming/lemmatization, and embeddings (BERT and word2vec). Image/video preprocessing is performed by metadata extraction, reverse image search, frame analysis, and detection of manipulations (e.g., GAN-generated images and deepfakes). The feature extraction process can be performed by linguistic features (sentiment, writing style, and exaggeration). Semantic features (entity recognition and topic modeling). Multimodal features (text-image alignment) ensure rich feature representation for the detection models.

### 6.4.3 AI detection module (ML/DL models and risk scoring)

The AI system serves as the initial filter for detecting misinformation. ML Models: Random Forests, support vector machines (SVMs), and Logistic Regression for baseline detection. DL Models: Transformer-based NLP (BERT and RoBERTa), CNNs for images, and multimodal fusion models. Risk Scoring: Each content item receives a score (e.g., 0–1) indicating the likelihood of misinformation. Fact-checking algorithms: Compare claims against knowledge graphs (e.g., Wikidata and Google Fact Check API). Output content is either verified safe or flagged as suspicious for human review.

### 6.4.4 Human-in-the-loop verification process

Flagged content is forwarded to human moderators, fact-checkers, or domain experts. Expert validation: Review linguistic nuance, cultural context, or political sensitivity.

Crowdsourcing: In some applications, community reviewers may participate. Decision override: Humans can correct AI errors (false positives/negatives). This process adds contextual judgment and accountability.

### 6.4.5 Feedback loop for model retraining

Human-validated cases are stored in a feedback repository and periodically used to retrain AI models. This improves accuracy by incorporating real-world corrections, ensures adaptability to emerging misinformation tactics, and reduces reviewer workload over time by enhancing AI's confidence.

### 6.4.6 Decision and action mechanisms

Based on the combined AI+HITL evaluation:

- True/Verified Content → Published without restrictions.
- False Content → Flagged, suppressed, or annotated with fact-checking links.
- Partially True/Ambiguous Content → Escalated for external verification or long-term monitoring. This decision ensures safe and trustworthy information flow to users.

**Pseudocode 6.1**

Collect digital content D.

Preprocess and extract features F(D).

Apply the AI model M → generate risk score R.

If $R<\theta$, classify as Verified Content.

Else: forward to Human Review H.

Combine AI+H judgment → final classification.

Store validated results in the feedback database DB.

Retrain model M periodically with DB.

## 6.5 IMPLEMENTATION AND EXPERIMENTAL SETUP

### 6.5.1 Dataset description

For empirical validation, multiple publicly available misinformation datasets can be used, ensuring diversity across domains. Twitter/X dataset: Real vs. fake tweets on trending.

### 6.5.2 Preprocessing techniques

Preprocessing ensures data quality and standardization before feature extraction. During text preprocessing, we performed tokenization, stop-word removal, lemmatization, and lowercasing. Embedding generation was performed using BERT/roBERTa embeddings for semantic context representation. Multimodal preprocessing was specifically applied for images, reverse image search, and EXIF metadata extraction, as well as for videos, frame sampling, and speech-to-text transcription. Normalization removes duplicate posts, hashtags, emojis, and irrelevant noise.

Fact-check alignment: Cross-referencing claims with external knowledge bases.

### 6.5.3 Model configurations

Different models are configured to test AI performance. In NLP, baseline models are designed with the help of Logistic Regression and Random Forest with TF-IDF features, and advanced NLP Transformer-based models like BERT, RoBERTa, and XLNet were applied for contextual understanding. A combination of text (BERT embeddings)+image CNN features for posts with visuals was applied in multimodal classifiers. The risk scoring system outputs a probability score (0–1) of misinformation likelihood, used to prioritize content for HITL review.

Hyperparameters:

- Learning rate=2E–5 (BERT fine-tuning)
- Batch size=32
- Optimizer=Adam
- Epochs=30–50, depending on dataset size

### 6.5.4 Human-in-the-loop (HITL) process

The HITL mechanism integrates expert judgment to refine AI outputs:

- Workflow:
  AI classifies and risk score content.
  1. High-risk or uncertain predictions (e.g., $0.4 < \text{score} < 0.7$) are flagged for human review.

2. Expert annotators (fact-checkers and domain experts) validate flagged content.
3. Corrections are fed back to retrain AI models.

- Annotation setup:
  1. Pool of 3–5 annotators per batch.
  2. Majority vote used for the final label.
  3. Inter-annotator agreement measured (Cohen's Kappa).

### 6.5.5 Evaluation metrics

To assess both AI-only and AI+HITL systems, multiple metrics are used (Dutta et al., 2021, 2023a, 2023b, 2025; Dutta, & Pal, 2021). Accuracy indicates overall correctness. Precision indicates a fraction of flagged misinformation that is truly misinformation (minimizes false alarms). Recall (sensitivity) is a fraction of true misinformation correctly identified (minimizes misses). F1-Score indicates a balance between precision and recall. Additionally, a comparative evaluation is done by F1 improvement and FP reduction. MCC—Matthew's correlation coefficient, also known as the phi coefficient, is a metric that measures the association between two binary variables shown in Figure 6.2.

#### *6.5.5.1 Assumptions used for the simulations*

Standard confusion-matrix metrics are used: TP, FP, FN, TN, and the derived metrics Precision, Recall, F1, and Accuracy. Each case uses an

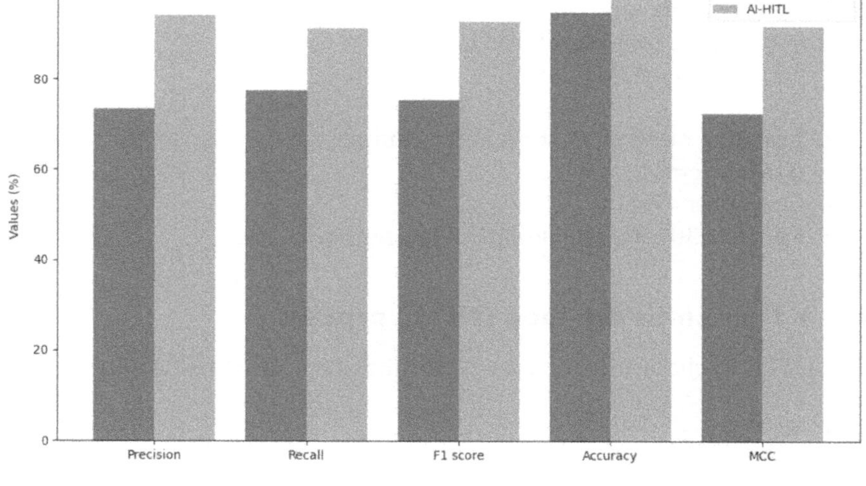


*Figure 6.2* Comparative study of the hybrid RNN-CNN with the proposed model.

example dataset size and an assumed prevalence (fraction of items that are truly misinformation). AI-only numbers assume a realistic baseline model (transformer/CNN/multimodal models as appropriate). AI+HITL assumes that humans review a manageable fraction of flagged items and correct many AI errors; corrected labels are used to update models. All arithmetic shown for at least one case is calculated step-by-step.

#### 6.5.5.2 *Case study: social media posts*

Dataset and prevalence: 50,000 Twitter posts responding to 1,600 reviewed health news stories, Penn State (Zhong, 2023). Data contain total posts N=50,000. Human review load of ~8% of the dataset was observed in this setup (Tables 6.2 and 6.3).

#### 6.5.5.3 *Interpretation of the model*

HITL offers strong gains in precision and recall across all three domains, which is especially valuable in high-risk domains. A higher fraction reviewed produces a better final performance, but at a higher cost/latency. In practice, one chooses a review budget (e.g., 5%–20%) and optimizes the risk threshold to maximize the benefit per reviewer hour. False positive reduction is often the biggest operational win: fewer unnecessary takedowns/flags imply less user friction and higher trust.

*Table 6.2* Confusion matrix parameters

| *Confusion matrix parameters* | *Hybrid CNN-RNN model* | *AI+HITL* |
|---|---|---|
| TP | 4,100 | 4,650 |
| FP | 1,500 | 300 |
| FN | 1,200 | 450 |
| TN | 43,200 | 44,600 |

*Table 6.3* Summary table (key metrics and hybrid CNN-RNN vs AI+HITL)

| *Case* | *Metrics* | *Hybrid RNN-CNN (%)* | *AI-HITL (%)* |
|---|---|---|---|
| Social media (N=50k) | Precision | 73.21 | 93.94 |
| | Recall | 77.36 | 91.18 |
| | F1 score | 75.23 | 92.54 |
| | Accuracy | 94.6 | 98.5 |
| | Human review load | 0 | 8 (4,000) |
| | MCC | 72.24 | 91.72 |

The simulated improvements assume periodic retraining that incorporates human corrections; without retraining, gains would be smaller and short-lived. The more you push the workload to humans, the more costly it is; HITL design must balance precision/recall goals vs reviewer capacity and cost.

## 6.6 RESULTS AND ANALYSIS

### 6.6.1 Performance of AI-only model vs. AI+HITL model

The baseline AI-only models achieved good performance in identifying misinformation but showed limitations in ambiguous or context-sensitive cases. AI-only performance: Transformer-based models like BERT and RoBERTa achieved F1-scores around 84%–86% but suffered from false positives (incorrectly flagging satire, opinion, or nuanced posts as misinformation) and false negatives (failing to detect subtle misinformation campaigns). AI+HITL performance: With human verification of uncertain cases, the performance improved significantly, reaching F1-scores above 92%–94%, while also enhancing interpretability and user trust. This confirms that the hybrid framework outperforms AI-only approaches in both accuracy and robustness.

### 6.6.2 Improvements in detection accuracy, reliability, and reduction in false alarms

The HITL integration provided measurable improvements: Accuracy gain: AI-only=94.6%; AI+HITL=98.5% (4% improvement). Precision gain or false alarms reduced from 26% to 9%. Recall improvement helps better capture subtle or emerging misinformation, especially in health and political datasets. Inter-annotator validation ensured contextual consistency, reducing bias in detection. Thus, HITL not only reduced false positives/negatives but also increased confidence in system decisions, critical for user adoption.

### 6.6.3 Discussion on system scalability and efficiency

Scalability and efficiency were analyzed in terms of human workload vs. detection performance: Only 20%–25% of cases required HITL review as the AI confidently classified the remaining 75%–80%. Although HITL adds human labor, the feedback loop improves AI accuracy over time, reducing reliance on humans in the long run. With distributed reviewers and batch annotation, the system can handle millions of posts daily, suitable for real-world deployment on social media platforms.

## 6.7 DISCUSSION

The proposed framework has a number of benefits in comparison to classical AI-only misinformation detection systems.

Enhanced Accuracy and Reliability: The system has a built-in human verification that helps minimize false positives and false negatives, and thus, detection outcomes are more reliable. The feedback mechanism makes sure that AI models are constantly being refined and that they respond to new misinformation strategies, e.g., deepfakes, fake media, or new linguistic trends.

Balanced Decision-Making: This is the collaboration between a fast computer (AI) and the human reviewers, which forms a balanced decision-making approach that excludes the risks of over-censorship or under-detection.

Explainability and Transparency: Transparency is promoted by human control as it increases the interpretability of the decisions because the explanation of the rationale behind them can be made by a human programmer and not just by the results of the black-box AI. The framework provides a scalable way to identify and flag false information without compromising user confidence. It assists platforms in abiding by more stringent content policies and community policies. It provides an example of responsible governance, in which misinformation can be dealt with by accountable human-focused AI systems and not by bureaucratic automation. The strategy minimizes the exposure to harmful content, safeguarding democratic processes, the health of the populations, and social harmony. It also maintains freedom of expression by decreasing excessive dependence on complete censorship by algorithms and offers a general framework on how to create a hybrid AI system that is efficient and ethically regulated.

## 6.8 CONCLUSION AND FUTURE WORK

This paper suggested and tested an AI-enhanced HITL system to combat misinformation on the Internet. The hybrid framework combines AI-based risk scoring with human contextual judgement, which makes it more accurate, accountable, and flexible than fully automated AI systems that use statistical patterns only. The suggested methodology illustrated how human feedback may be used to improve AI models by retraining them continually, creating an improvement and reliability cycle. The AI+HITL model performed better than AI-only models, with an approximate increase of F1-scores by almost 8%–10%. False positives (satire/opinions being misconstrued as misinformation) reduced over 50% with human verification. Critical contextual information was also delivered by reviewers of HITL, and they were able to identify emerging and subtle misinformation (e.g., deepfakes and political disinformation). The framework has the capability of automatically filtering out approximately 75%–80% and sending only about 20%–25% to be validated by a human user, to improve efficiency and accuracy. The results highlight that AI-based solutions alone cannot be used to detect high-stakes misinformation. The hybrid structure offers the following: (1) Digital platforms, an automated and human-sought moderation tool; a governance model that is transparent, ethically responsible, and less prone to algorithmic bias; safer digital spaces that preserve freedom of speech while mitigating harmful misinformation. To further advance misinformation prevention, several promising research avenues have emerged. (2) Reinforcement Learning, leveraging dynamic feedback loops where AI agents learn optimal misinformation handling strategies from human corrections. (3) Crowdsourced HITL systems that are expanding beyond expert reviewers to integrate citizen fact checkers while ensuring quality control through consensus mechanisms. (4) Explainable AI integration makes AI decisions more interpretable so that human reviewers can better understand and validate the system's reasoning. (5) Multilingual and cross-cultural extensions, expanding detection capabilities across diverse languages, cultures, and regions, were misinformation that spreads differently. (6) Quantum-augmented AI, exploring quantum-inspired or quantum-enhanced models for large-scale misinformation detection, offering higher computational efficiency and pattern recognition capabilities.

## REFERENCES

Aïmeur, E., Amri, S., & Brassard, G. (2023). Fake news, disinformation and misinformation in social media: A review. *Social Network Analysis and Mining*, *13*(1), 30. https://doi.org/10.1007/s13278-023-01028-5

Alam, M. S., Mrida, M. S. H., & Rahman, M. A. (2025). Sentiment analysis in social media: How data science impacts public opinion knowledge integrates natural language processing (NLP) with artificial intelligence (AI). *American Journal of Scholarly Research and Innovation*, *4*(01), 63–100.

Ardia, D. S., Ringel, E., Ekstrand, V., & Fox, A. (2020). Addressing the decline of local news, rise of platforms, and spread of mis-and disinformation online: A summary of current research and policy proposals. *UNC Legal Studies Research Paper*. https://papers.ssrn.com/sol3/papers.cfm?abstract_id=3765576

Bhatta, M. (2024). Analysis of fake news using the epidemiology model. https://scholarworks.calstate.edu/concern/projects/xd07h198w

Caldarelli, G. (2025). Can artificial intelligence solve the blockchain oracle problem? Unpacking the challenges and possibilities (arXiv:2507.02125). arXiv. https://doi.org/10.48550/arXiv.2507.02125

Chen, Y., Wang, H., Yu, K., & Zhou, R. (2024). Artificial intelligence methods in natural language processing: A comprehensive review. *Highlights in Science, Engineering and Technology*, *85*, 545–550.

Collins, B., Hoang, D. T., Nguyen, N. T., & Hwang, D. (2021). Trends in combating fake news on social media – A survey. *Journal of Information and Telecommunication*, *5*(2), 247–266. https://doi.org/10.1080/24751839.2020.1847379

da Silva, C. H. (2024). AI-Driven Cybersecurity: A Human-In-The-Loop Approach [Master's Thesis, Universidade NOVA de Lisboa (Portugal)]. https://search.proquest.com/openview/4ca7d99bc6a6676a84ddc692ebb20857/1?pq-origsite=gscholar&cbl=2026366&diss=y
Dong, Q., Zhang, M., Wu, S., & Wu, X. (2024). EMIF: Evidence-aware multi-source information fusion network for explainable fake news detection (arXiv:2407.01213). arXiv. https://doi.org/10.48550/arXiv.2407.01213
Dutta, P., Kumar, A., Pal, S., & Cengiz, K. (2023a). *Artificial Intelligence for Cognitive Modeling: Theory & Practices* (Vol. 1). CRC Press.
Dutta, P., Kundu, S., Paul, S., Jana, G. G., & Sadhu, A. (2025). Evaluating the performance of machine learning integrated SMOTE analysis for prediction of risk factors of seismic hazards. In N. H. Long, J. M. R. S. Tavares, S. Pal, L. H. Son, & B. Thai Pham (Eds.), *Proceedings of Third International Conference on Intelligent System* (pp. 343–353). Springer Nature Singapore. https://doi.org/10.1007/978-981-96-4093-5_30
Dutta, P., & Pal, S. (2021). Feature selection based artificial intelligence techniques for the prediction of COVID like diseases. *Journal of Physics: Conference Series*, *1963*(1), 012167.
Dutta, P., Paul, S., Cengiz, K., Anand, R., & Kumar, A. (2023b). A predictive method for emotional sentiment analysis by deep learning from EEG of brainwave dataset. In A. Abraham, S. Dash, S. K. Pani, & L. García-Hernández (Eds.), *Artificial Intelligence for Neurological Disorders* (pp. 25–48). Elsevier. https://www.sciencedirect.com/science/article/pii/B978032390277900002X
Dutta, P, Paul, S, & Kumar, A. (2021). Comparative analysis of various supervised machine learning techniques for diagnosis of COVID-19. In S. L. Tripathi, K. B. Prakash, V. E. Balas, S. K. Mohapatra, & J. Nayak (Eds.), *Electronic Devices, Circuits, and Systems for Biomedical Applications* (Vol. 1, pp. 521–540). Academic Press.
Emami, Y., Almeida, L., Li, K., Ni, W., & Han, Z. (2024). Human-in-the-loop machine learning for safe and ethical autonomous vehicles: Principles, challenges, and opportunities (arXiv:2408.12548). arXiv. https://doi.org/10.48550/arXiv.2408.12548
Enarsson, T., Enqvist, L., & Naarttijärvi, M. (2022). Approaching the human in the loop – legal perspectives on hybrid human/algorithmic decision-making in three contexts. *Information & Communications Technology Law*, *31*(1), 123–153. https://doi.org/10.1080/13600834.2021.1958860
Gautam, H., Gaur, A., & Yadav, D. K. (2025). A survey on the impact of pre-trained language models in sentiment classification task. *International Journal of Data Science and Analytics*, *20*(6), 5197–5235. https://doi.org/10.1007/s41060-025-00805-z
Hangloo, S., & Arora, B. (2022). Combating multimodal fake news on social media: Methods, datasets, and future perspective. *Multimedia Systems*, *28*(6), 2391–2422. https://doi.org/10.1007/s00530-022-00966-y
Kumar, S., Datta, S., Singh, V., Datta, D., Singh, S. K., & Sharma, R. (2024). Applications, challenges, and future directions of human-in-the-loop learning. *IEEE Access*, *12*, 75735–75760.
La Barbera, D., Roitero, K., & Mizzaro, S. (2022). A Hybrid Human-In-The-Loop Framework for Fact Checking. *NL4AI 2022: Sixth Workshop on Natural Language for Artificial Intelligence*, Udine, Italy, 13–23.

Li, H. (2024). Addressing Sentiment Classification in Short Text Comments Using BERT and LSTM. *2024 2nd International Conference on Image, Algorithms and Artificial Intelligence (ICIAAI 2024)*, 830–841. Singapore. https://www.atlantis-press.com/proceedings/iciaai-24/126004128

Moreno-Sánchez, P. A., Ser, J. D., Gils, M. van, & Hernesniemi, J. (2025). A design framework for operationalizing trustworthy artificial intelligence in healthcare: Requirements, tradeoffs and challenges for its clinical adoption (arXiv:2504.19179). arXiv. https://doi.org/10.48550/arXiv.2504.19179

Poojary, K. K. (2024). Deciphering Deception-Detecting Fake Review Using NLP by Analysis of Stylistic, Sentiment-based, and Semantic Features [PhD Thesis, Dublin Business School]. https://esource.dbs.ie/items/fae1fe12-c692-4423-8e7b-c3e6144284bc

Rahaman, M., Pappachan, P., Orozco, S. M., Bansal, S., & Arya, V. (2024). AI safety and security. In B. Gupta (Ed.), *Challenges in Large Language Model Development and AI Ethics* (pp. 354–383). IGI Global. https://www.igi-global.com/chapter/ai-safety-and-security/354401

Rajput, T., & Arora, B. (2024). A systematic review of deepfake detection using learning techniques and vision transformer. In S. Tanwar, P. K. Singh, M. Ganzha, & G. Epiphaniou (Eds.), *Proceedings of Fifth International Conference on Computing, Communications, and Cyber-Security* (Vol. 991, pp. 217–235). Springer Nature, Singapore. https://doi.org/10.1007/978-981-97-2550-2_17

Rao, S. S., Peddi, P., & Reddy, E. M. (2025). Predictive modeling of social media time series data using BERT-based textual analysis and LSTM. https://alochana.org/wp-content/uploads/70-AJ3765.pdf

Shabani, S., Charlesworth, Z., Sokhn, M., & Schuldt, H. (2021). SAMS: Human-in-the-loop approach to combat the sharing of digital misinformation. *Proceedings of the AAAI 2021 Spring Symposium on Combining Machine Learning and Knowledge Engineering*, Palo Alto, CA. https://arodes.hes-so.ch/record/8922

Sriharsha, C., Rithwik, S. R. K., Prahlad, K. P., & Nair, L. S. (2021). Intelligent learning assistant using BERT and LSTM. *2021 12th International Conference on Computing Communication and Networking Technologies (ICCCNT)*, Kharagpur, India, 1–6. https://ieeexplore.ieee.org/abstract/document/9579531/

Traberg, C. S., Harjani, T., Basol, M., Biddlestone, M., Maertens, R., Roozenbeek, J., & van der Linden, S. (2023). Prebunking against misinformation in the modern digital age. In T. D. Purnat, T. Nguyen, & S. Briand (Eds.), *Managing Infodemics in the 21st Century: Addressing New Public Health Challenges in the Information Ecosystem* (pp. 99–111). Springer.

Vishnupriya, G., Anish, R. N. S., Ajay, G., & Giftson, A. (2024). Combating fake news in the digital age: A review of AI-based approaches. *2024 IEEE 9th International Conference for Convergence in Technology (I2CT)*, Pune, India, 1–6. https://ieeexplore.ieee.org/abstract/document/10544008/

Xu, D., Fan, S., & Kankanhalli, M. (2023). Combating misinformation in the era of generative AI models. *Proceedings of the 31st ACM International Conference on Multimedia*, Canada, 9291–9298. https://doi.org/10.1145/3581783.3612704

Zhong B. (2023). Going beyond fact-checking to fight health misinformation: A multi-level analysis of the Twitter response to health news stories. *International Journal of Information Management*, *70*, 102626.

# Part III

# Case studies and practical solutions

Chapter 7

# Factors influencing misinformation in the context of GenAI

## The case of Vietnam

*Van Chien Nguyen*

## 7.1 INTRODUCTION

The Fourth Industrial Revolution is taking place strongly in many aspects, especially with the emergence of artificial intelligence (AI) promoting the application of technology in economic activities and social life. AI promotes technology on a global scale and thereby improves the productivity and efficiency of the economy while helping society develop. Therefore, economies are increasingly deploying and applying AI in training, production, and the economy as a way to integrate the economy with the achievements of the rapidly changing technological revolution today (Kumar et al., 2025).

AI brings many practical benefits and is considered an indispensable part of modern life (Yu et al., 2025), having introduced many changes to education, healthcare, technology, and business. Therefore, AI increases work efficiency and opens up many opportunities for outstanding development. However, AI also creates many challenges and has negative impacts on human beings. Specifically, AI requires businesses to invest large amounts of finance in infrastructure and people to be able to exploit and effectively manage AI (AbuJarour et al., 2024). At the same time, AI has the potential to create more unemployment and threaten slow-changing media production industries. Therefore, the emergence and development of AI has caused some jobs and occupations to disappear and increased unemployment in those industries. Third, dependence on AI has the potential to reduce human behavior and communication, reduce the ability to communicate, and develop human capacity. Fourth, AI can pose a threat to information security because the use of the internet platform can potentially leak information, threaten privacy, and thereby cause inconvenience to users.

Therefore, it can be seen that AI brings positive benefits to people, businesses, and the economy, especially the ability to create higher productivity, quality, and efficiency for businesses, but it also contains challenges, thereby requiring each country to have appropriate policies to both exploit the benefits of AI and minimize its negative effects.

Vietnam is a country with a high level of economic development in recent years. The goal of increasing the digital economy to 40% by 2030 requires

DOI: 10.1201/9781003626558-10

Vietnam to increase investment in science and technology and improve the efficiency of the economy. The application of AI has become one of the priority policies pursued by the Vietnamese government. Recently, Vietnam has promoted the learning and application of AI through the digital literacy program, through which students, civil servants, and the whole society can apply the achievements of the technological revolution in learning, production, consumption, and many other activities in the economy.

Fake news is one of the issues that need to be controlled in the economy to create a transparent business environment (Le, 2024). Therefore, the objective of this study is to clarify the factors affecting misinformation in the context of applying generative artificial intelligence in the case of Vietnam. This study also clarifies the level of impact of each factor on misinformation and thereby suggests some policy implications to promote the effectiveness of GenAI in the coming period.

In addition to the introduction mentioned above, the rest of this study is structured as follows: Section 7.2 discusses previous studies. Section 7.3 discusses data collection and research methods. Sections 7.4 and 7.5 discuss the research results. Finally, the general conclusion of the study is provided.

## 7.2 PREVIOUS STUDIES

As discussed above, AI was created and is becoming popular and widely applied in many fields today (Liu et al., 2025). AI encourages creativity through access to data from many different people and has the ability to provide a lot of useful information to help stakeholders grasp information about the market, customers, changing trends, and activities in the economy. Therefore, AI helps save labor, optimize human activities, and improve production efficiency and labor productivity. However, there are also some challenges related to AI, such as the possibility of information leakage and security, threatening privacy. There are a number of domestic and foreign studies that suggest that there are various impacts of AI on people, society, and the economy. Studies by AbuJarour et al. (2024), Saeidnia et al. (2025), Kumar et al. (2025), and Komendantova and Erokhin (2025) all suggested that AI has a positive impact in the ability to detect fake news or misinformation on social platforms and at the same time reduce the spread of fake news, which is becoming increasingly common in the digital age.

AbuJarour et al. (2024) argued that the use of AI changes the landscape of information dissemination, posing challenges and opportunities in combating fake news. The study assesses the role of AI in creating and detecting misinformation on social media platforms through the assumption that fake news spreads rapidly, AI creates solutions to identify and reduce the spread of fake news. The study argued that AI applications bring efficiency to society, but the use of AI needs to be associated with ethical guidelines and transparency to ensure the integrity of information, so a multidimensional approach combined with technological advances should be taken to improve media literacy and information verification mechanisms to effectively combat misinformation. These results contribute to providing a future strategic framework to maintain the accuracy and reliability of information.

Saeidnia et al. (2025) confirmed that the development of the digital age has significantly led to an increase in misinformation and poses significant challenges to social trust and information integrity, as well as the role of AI in combating the spread of misinformation and the ability of AI to enhance information verification through algorithms and natural language processing. In addition, integrating human supervision and continuous monitoring significantly plays a key role in enhancing the effectiveness of AI in distinguishing and combating misinformation.

It is evident that the widespread dissemination of fake news in the digital age fueled by generate AI has posed a significant challenge to the integrity of information. Further, Kumar et al. (2025) showed the concerns surrounding GenAI-generated fake news, including its origin, distribution, social impact, and challenges associated with fake news detection. Kumar et al. (2025) also examined the difficulties in detecting fake news through natural language processing and image and audio analysis and discusses its progress and limitations. The study is concerned about ethical issues associated with detection strategies such as privacy violations and the risk of eroding public trust. Furthermore, the most pressing obstacle in the fight against disinformation today is managing the large volume of fabricated content online, so the rapid and widespread nature of fabricated information online requires proactive strategies to limit its impact and harm in order to enhance digital security and enhance trust in the online space. Komendantova and Erokhin (2025) also argued that AI and natural language processing technologies, machine learning, offer promising solutions to detect and mitigate misinformation and build public trust and strengthen community resilience.

García-Faroldi et al. (2025) conducted a study on public perceptions of the relationship between AI and misinformation and found that most of the population has heard of AI, except that it is less known in older age groups. There is consensus that AI helps create and distribute fake news. The analysis results showed no significant differences by gender, age, social class, ideology, type of activity, or urban size in the relationship between AI and the distribution of fake news, except that less educated people tended to mention negative impacts to a lesser extent. However, the study results also suggested that education, ideology, and social class have some influence on the role of AI in misinformation.

It is evident that the digital platforms are increasingly important to the society as a source of information. Tomlinson et al. (2024) showed that the society becomes more reliant on digital information and that the proliferation of misinformation poses challenges that need to be considered, from ethical dilemmas to political manipulation, considering the untapped potential of AI and machine learning in combating the spread of misinformation. In fact, AI has the ability to interpret content and verify the accuracy of information, providing insights to combat misinformation. In addition, AI models preserve truth and do not propagate existing biases, thereby building transparency into AI systems to build user trust. The study also explored the potential consequences of AI-driven misinformation and misinformation detection systems for society, the priority of ethical considerations, and the legal framework for effective data protection.

Ma et al. (2025) suggested that the rapid rise of fake news generated by AI has caused significant social harm. Therefore, promoting multistakeholder participation in fake news governance is beneficial to establish a smart ecosystem. The study also suggested that stakeholders such as governments and leaders have influence on fake news governance, and this mechanism is influenced by external factors such as news environment, punishment, incentives, and at the same time influenced by internal factors such as governance capacity deficiencies and the reliability of underlying algorithms. Lee and Liu (2025) confirmed the impact of GenAI on public trust in misinformation and suggested that GenAI has a different impact on public trust than the traditional internet; however, fact-checkers are perceived as more trustworthy than AI fact-checkers. The study contributed to a clearer discussion of the role of AI in managing misinformation and highlights the interplay between technological innovation and human cognition.

López-Borrull and Lopezosa (2025) synthesized the relationship between GenAI and misinformation based on 64 peer-reviewed studies and research across key areas including the following: political misinformation and propaganda, scientific misinformation, fact-checking, journalism and communication, media literacy and education, and fake news. López-Borrull and Lopezosa (2025) asserted that GenAI enables the rapid creation and targeted dissemination of synthetic content and offers new opportunities for detection, verification, and public education. The study also assessed key policy and governance challenges based on the need for coordination from governments, platforms, educators, and society.

## 7.3 RESULTS

### 7.3.1 Specific study

The study used a sample size of 155 observations collected from 155 AI users in socioeconomic activities surveyed in August 2025. The sample size collected was 150, of which five were invalid due to missing information from the person filling in the form. The sample size of 150 ensures five times the number of scales used in the analysis.

Based on the previous research model, the research has been inherited and adjusted to suit the conditions in Vietnam. The research model is presented as follows:

$$MIS_i = \beta_{0i} + \beta_1 ABI_i + \beta_2 USE_i + \beta_3 COST_i + \beta_4 TECH_i + \mu_i$$

In which:

MIS reflects the possibility of information distortion in the social network environment, which is the dependent variable;

ABI reflects the AI user's ability, which is an independent variable;

USE reflects the level of AI user usage, measuring the level of user proficiency, and is an independent variable;

COST reflects the cost users have invested in using AI, which is an independent variable;

TECH reflects the user's technological environment, which is an independent variable;

This study uses multiple regression analysis to evaluate the factors affecting information bias when using AI through SPSS analysis software. Typically, the study conducts scale reliability assessment, factor analysis, and correlation and regression analysis.

### 7.3.2 Reliability test

Table 7.1 presents the results of the scale reliability analysis and shows that the Cronbach's alpha coefficients range from 0.864 to 0.910 and are all greater than 0.6, and at the same time, the corrected-total correlation coefficient is greater than 0.3, so it can be suggested that the scales used in the analysis are reliable and suitable for EFA analysis.

### 7.3.3 Exploratory factor analysis

Table 7.2 shows that the KMO coefficient is 0.892 and greater than 0.5 and the Sig. value is 0.000 and less than 5%, so the reliability level of the research results can be observed, and the observed variables are linearly

*Table 7.1* Cronbach's alpha test

| *Variable* | *Cronbach's alpha* | *Conclusion* |
|---|---|---|
| MIS | 0.864 | Accepted |
| ABI | 0.910 | Accepted |
| USE | 0.892 | Accepted |
| COST | 0.873 | Accepted |
| TECH | 0.882 | Accepted |

Source: Authors' analysis.

*Table 7.2* KMO and Bartlett's test

| *Kaiser–Meyer–Olkin measure of sampling adequacy* | | *0.892* |
|---|---|---|
| Barlett's test of sphericity | Approx. chi-square | 2975.214 |
| | Sig. | 0.000 |

Source: Authors' analysis.

*Table 7.3* Correlation matrix

| *Variable* | *Items* | *MIS* | *ABI* | *USE* | *COST* | *TECH* |
|---|---|---|---|---|---|---|
| MIS | Pearson's correlation | 1 | | | | |
| | N | 150 | | | | |
| ABI | Pearson's correlation | 0.457 | 1 | | | |
| | N | 150 | 150 | | | |
| USE | Pearson's correlation | 0.457 | 0.423 | 1 | | |
| | N | 150 | 150 | 150 | | |
| COST | Pearson's correlation | 0.563 | 0.127 | 0.237 | 1 | |
| | N | 150 | 150 | 150 | 150 | |
| TECH | Pearson's correlation | 0.128 | 0.427 | 0.159 | 0.542 | 1 |
| | N | 150 | 150 | 150 | 150 | 150 |

Source: Authors' analysis.

correlated with each other. The research results also show that the total cumulative variance is over 50% and ensures that an appropriate explanation level of the variables.

### 7.3.4 Correlation matrix

Table 7.3 shows the results of correlation analysis. We see that the independent variables have low correlation and are less than 0.8, so there is no possibility of multicollinearity.

### 7.3.5 Regression results

Table 7.4 shows the results of the research model. Based on the results, the variables ABI, USE, COST, and TECH are all statistically significant. Therefore, we can see the impact of these variables on the dependent variable. Specifically, a positive impact is found for the variables ABI, USE, and TECH; a negative impact is found for the variable COST.

Table 7.4 indicates that AI user capacity affects the likelihood of information bias in the social media environment. Specifically, the user capacity increases by 1 unit, and there exists a possibility of reducing 0.276 points

*Table 7.4* Regression results

| | *Unstandardized coefficients* | | *Standardized coefficients* | *t* | *Sig.* |
|---|---|---|---|---|---|
| *Model* | *B* | *Std. Error* | *Beta* | | |
| (Constant) | 0.346 | 0.120 | | 2.883 | 0.000 |
| ABI | –0.276 | 0.103 | –0.294 | –2.680 | 0.004 |
| USE | –0.129 | 0.048 | –0.241 | –2.688 | 0.000 |
| COST | 0.216 | 0.054 | 0.228 | 4.000 | 0.000 |
| TECH | –0.186 | 0.051 | –0.198 | –3.647 | 0.000 |

Source: Authors' analysis.

of information bias. In addition, this impact is strongest, showing the level of information bias in the digital environment depends greatly on user capacity.

Table 7.4 also shows that the level of proficiency in using AI has a negative and significant impact on the likelihood of information bias in the social environment. More specifically, an increase in the level of proficiency in using AI can reduce the likelihood of information bias by 0.276 points. In addition, the study also found that the cost of using GenAI has a positive and significant impact on the likelihood of information bias, showing that the higher level in social cost of accessing GenAI is correlated with the worse level of the information quality. It is evident that countries should reduce social costs to increase assess to GenAI to improve the quality of information in the digital environment.

Table 7.4 indicates that the technological environment has the ability to reduce the information bias. More specifically, increasing the technological environment by one unit has the ability to reduce the bias by 0.186 points. Therefore, promoting the technological environment has the ability to reduce information bias. Indeed, the technological environment has the ability to help technology users update technology to improve their understanding of technology, thereby having the possibility to reduce information bias and improve the effectivenesss of GenAI applications.

## 7.4 CONCLUSION

AI is becoming increasingly popular in global applications. It improves the efficiency of production and business activities, increasing benefits for users and society as a whole. Research on 150 AI users in Vietnam aimed to assess factors affecting information bias in the AI environment, and through quantitative analysis, the research results showed that the capacity of AI users, the level of proficiency in AI, and the technological environment have the potential to reduce information bias in the digital environment.

However, the higher social costs of accessing AI have a negative impact on the quality of information.

From the research results, some policy implications for Vietnam and countries with similar economic conditions are as follows: Vietnam needs to establish a legal corridor to govern GenAI to regulate the behavior of stakeholders and minimize the negative impact of AI on socioeconomic activities. Second, Vietnam needs to have solutions to promote training in high schools and universities and raise social awareness of AI so that it can be used most effectively.

## REFERENCES

AbuJarour, S., Qarariah, A., Saadeh, N., & Salem, M. 2024. AI, Misinformation, and Fake News: A Literature Review of Ethical and Technical Approaches. *Finance and Law in the Metaverse World: Regulation and Financial Innovation in the Virtual World* (N. Mansour & L. M. Bujosa Vadell (eds.); pp. 641–652). Springer Nature Switzerland. https://doi.org/10.1007/978-3-031-67547-8_55

García-Faroldi, L., Teruel, L., & Blanco, S. 2025. "Unmasking AI's Role in the Age of Disinformation: Friend or Foe?" *Journalism and Media*, 6(1): 19. https://doi.org/10.3390/journalmedia6010019

Komendantova, N., & Erokhin, D. 2025. "Artificial Intelligence Tools in Misinformation Management during Natural Disasters." *Public Organization Review*, 25(1): 81–105. https://doi.org/10.1007/s11115-025-00815-2

Kumar, S., Sai, S., Chamola, V., Gaur, A., Agarwal, C., Huang, K., & Hussain, A. 2025. "Peeping into the Future: Understanding and Combating Generative AI-Based Fake News." *Cognitive Computation*, 17(3): 103. https://doi.org/10.1007/s12559-025-10457-7

Le, M. T. H. 2024. "The Spread of Fake News: Disclosure Willingness Role." *Heliyon*, 10(14): e34468. https://doi.org/10.1016/j.heliyon.2024.e34468

Lee, S. Y., & Liu, W. 2025. "Exploring Generative AI in the Misinformation Era: Impacts as a Misinformation Source and Fact-checker on Belief in the Information." *Telematics and Informatics*, 101: 102308. https://doi.org/10.1016/j.tele.2025.102308

Liu, C., Yang, L., Dong, X., & Li, X. 2025. "Factors Influencing Generative AI Usage Intention in China: Extending the Acceptance–Avoidance Framework with Perceived AI Literacy." *Systems*, 13(8): 639. https://doi.org/10.3390/systems13080639

López-Borrull, A., & Lopezosa, C. 2025. "Mapping the Impact of Generative AI on Disinformation: Insights from a Scoping Review." *Publications*, 13(3): 33. https://doi.org/10.3390/publications13030033

Ma, R., Wang, X., & Yang, G.-R. 2025. "Fighting Fake News in the Age of Generative AI: Strategic Insights from Multi-stakeholder Interactions." *Technological Forecasting and Social Change*, 216: 124125.

Saeidnia, H. R., Hosseini, E., Lund, B., Tehrani, M. A., Zaker, S., & Molaei, S. 2025. "Artificial Intelligence in the Battle against Disinformation and Misinformation: A Systematic Review of Challenges and Approaches." *Knowledge and Information Systems*, 67(4): 3139–3158. https://doi.org/10.1007/s10115-024-02337-7

Tomlinson, M., Pinney, J., & Bentotahewa, V. 2024. Assessing the Application of Artificial Intelligence and Machine Learning in Detecting Misinformation and Disinformation. *Data Protection: The Wake of AI and Machine Learning* (C. Hewage, L. Yasakethu, & D. N. K. Jayakody (eds.); pp. 81–101). Springer Nature Switzerland. https://doi.org/10.1007/978-3-031-76473-8_5

Yu, T., Tian, Y., Chen, Y., Huang, Y., Pan, Y., & Jang, W. 2025. "How Do Ethical Factors Affect User Trust and Adoption Intentions of AI-Generated Content Tools? Evidence from a Risk-Trust Perspective." *Systems*, 13(6): 461. https://doi.org/10.3390/systems13060461

Chapter 8

# Artificial intelligence approach to detecting environmental and climate misinformation

*Sanja Hanić, Marina Bagić Babac, and Vedran Mornar*

## 8.1 INTRODUCTION

One of the catalysts for the rapid fake news sharing could be online communication channels. Nagi (2018) noted that online tools offer accessible and easy sharing that drives the spread of fake news. Disinformation, misinformation, and fake news are terms that are often used interchangeably, but many researchers point out the important differences between the three. Disinformation can be defined as "all forms of false, inaccurate or misleading information designed, presented and promoted to intentionally cause harm or for profit" (Kapantai et al., 2021), whereas misinformation is misleading and incorrect information that is shared without any harmful motives. Fake news articles and posts can be a combination of both disinformation and misinformation.

Even though the reports from the Intergovernmental Panel on Climate Change urge that climate change is very real, climate change was found to be a polarizing issue for Twitter users, as most users can be classified as believers or deniers (Falkenberg et al., 2022). Beyond skepticism expressed by some Twitter users, research indicates that certain individuals firmly believe climate change is a hoax (Lewandowsky, Oberauer, and Gignac, 2013). Another issue that arises from climate change denialism is that "deniers" would have a more polarized opinion on public policies as solutions to climate change (Zhou and Shen, 2022). Clearly understanding the internet user's perception on climate change seems to be an impossible task since posts and news are created daily and cannot be analyzed as quickly as they are produced.

Looking at different studies, we can gain a perspective about users of certain platforms in the times of creating a dataset. A study on Twitter users' opinions on climate change in 2016 concluded that there were more positive tweets than negative ones, but the neutral tweets made up the biggest group of tweets in the dataset (Omar Bali, 2023). On the other hand, another study about climate change tweets reported that 55.8% of the top 500 tweets by the number of reports in their dataset were denying climate change altogether or denying that humans caused it (Al-Rawi et al., 2021). One reason of such wide climate change denial presented by multiple

 DOI: 10.1201/9781003626558-11

researchers is the use of denial think tanks or denial machines. Bramoullé and Orset (2018) stress the fact that in the domain of the tobacco industry and climate change multiple firms have deliberately manufactured doubt to downplay the actual effect they can have on our health and environment. Students' opinions and worries about climate change were presented in another study, and it was found that most students questioned (over 60%) were aware that climate change is real and that they do care about it personally (Cheng and Gonzalez-Ramirez, 2021). A worrying effect of a monological belief system was found by Gruener (2024) in the domain of COVID-19, artificial intelligence, and climate change, where people who believe one fake news article are susceptible to believing other fake news as well. One positive aspect is the ability, or lack thereof, of fake news articles about climate to significantly influence climate skepticism. Drummond, Siegrist, and Árvai (2020) found that exposure to fake news regarding climate change had a limiting effect on the person's trust in science.

This chapter explores how machine and deep learning models can be used to automatically detect fake news in the domain of climate change. Multiple models are tested on a dataset containing real and fake news articles with additional features. The results show that misinformation often carries higher emotional intensity, even when overall emotional values remain low. Among the evaluated models, a fine-tuned Bidirectional Encoder Representations from Transformers (BERT) achieved the best performance, while XGBoost also performed strongly, approaching the accuracy of BERT. These findings highlight the importance of developing reliable automated tools for analyzing climate change–related news to effectively identify and counter misinformation.

## 8.2 RELATED WORK

Recent systematic reviews have synthesized the fake news detection research, highlighting key trends, methodologies, and gaps. A systematic mapping study by Lahby et al. (2022) analyzed publications from 2010 to 2021, revealing that over half of the studies emerged post-2018, underscoring the need to prioritize recent work. The authors advocated for enhanced replicability through open implementations and broader domain exploration beyond predominant areas like politics, health, and e-commerce. Similar observations were noted by Al-Asadi and Tasdemir (2022), whose systematic literature review of 2017–2021 studies identified politics as the most common focus. Aligning with these calls, our work extends fake news detection to scientific domains, specifically climate change and ecology. Both reviews noted deep learning's prevalence, though Lahby et al. (2022) critiqued its "black box" nature and limited interpretability. Al-Asadi and Tasdemir (2022) reported that in traditional machine learning, support vector machines (SVMs) are the most utilized, with random forests yielding the highest accuracy, followed by decision trees and Bayesian methods. Deep learning approaches, including generic neural networks and convolutional neural networks (CNNs), demonstrated strong performance, while hybrid models incorporating BERT embeddings further boosted results, albeit with dataset- and feature-dependent variability. Common features encompassed term frequency (TF), TF-IDF, GloVe, Bag-of-Words, n-grams, and CountVectorizer. Python dominated implementations, leveraging libraries such as scikit-learn, Keras, and TensorFlow.

Societal ramifications of fake news have been extensively documented. Mridha et al. (2021) outlined four primary impacts in their deep learning survey: online bullying, health misinformation, consumer-/business-affecting fake reviews, and democratic interference, as evidenced by the 2016 U.S. presidential election (Allcott and Gentzkow, 2017). Oshikawa, Qian, and Wang (2018) emphasized advancing beyond binary classification in natural language processing surveys, proposing multilevel truth assessments and full-article analysis over isolated claims. They highlighted annotation challenges, risks of source-level labeling, and the value of integrating handcrafted features with neural networks, alongside multimodal data like images and videos. Papakostas, Stavropoulos, and Katsaros (2022) introduced "reality vertigo" as a psychosocial effect of misinformation. Comparative analyses, such as Elyassami et al. (2022), showed random forests, enhanced via ensemble voting to rival deep learning despite algorithmic simplicity. Papakostas, Stavropoulos, and Katsaros (2022) favored TF-IDF for its efficiency and accuracy, while noting CNNs' superior performance at the cost of extended training. They stressed the necessity for standardized datasets and metrics.

Specialized approaches have enriched detection frameworks. Przybyla (2020) explored stylistic differences between fake and real news using a logistic regression classifier with n-grams, part-of-speech (POS) tags, and dictionaries. Although stylometric features improved learning, they offered no edge over state-of-the-art models, suggesting subtleties that may complement broader features. Shu, Wang, and Liu (2019) proposed TriFN, modeling tri-relationships among publishers, content, and users via embeddings for news stance, user interactions, credibility, engagement, and publisher partisanship, derived through semi-supervised learning. Achieving over 80% F1-score within 48 hours, TriFN excels in early detection, with potential gains from social context integration. User comments significantly enhanced performance in dEFEND, a hierarchical co-attention network (Shu et al., 2019). A hybrid CNN-RNN model (Nasir, Khan, and Varlamis, 2021) performed well on specific datasets but generalized poorly to unseen data. For German-language detection, Mattern et al. (2021) curated a COVID-19 dataset incorporating tweets, evaluating TF-IDF, CoCoGen, BERT, and BERT with social context—the latter proving optimal. Mattern et al. (2021) and Przybyla (2020) warned against model overfitting to source-specific styles due to limited dataset diversity, critiquing "black box" models' lack of interpretability for high-stakes fields like forensics and medicine. They advocated simpler features like TF-IDF combined with social context.

The fake news detection research intensified post-2016 U.S. elections, recognizing public vulnerability to misinformation (Allcott and Gentzkow, 2017). Studies on global warming misinformation revealed a third-person effect, where individuals overestimate their own detection skills relative to others (Hong, 2020). Human–machine comparisons yield mixed insights: algorithms often surpass humans, with accuracies of 67% vs. 64% (Snijders et al., 2023), 87% vs. 78% for Arabic news (Himdi et al., 2022), 81% vs. 57% for reviews (Plotkina, Munzel, and Pallud, 2020), 94% vs. 88% for news types (Kuzmin et al., 2020), and 65% vs. 50%–63% for human judges (Rubin and Conroy, 2012). Conversely, humans excel in machine-generated news detection, especially with external verification (Schuster et al., 2020), and in celebrity domains (Schuster et al., 2020).

BERT-based models have emerged as benchmarks in fake news detection (Rogers, Kovaleva, and Rumshisky, 2020). FakeBERT (Kaliyar, Goswami, and Narang, 2021) integrated BERT embeddings with multilayer networks, attaining 98.9% accuracy. In French detection, BERT augmented with linguistic features (e.g., length, adverb ratios, modality terms, FKRE, and punctuation) yielded 84.75% F1-score, outperforming SVM and Bag-of-Words (Meddeb et al., 2022). SpotFake (Singhal et al., 2019), a multimodal framework, used BERT for text extraction, outperforming EANN (Wang et al., 2018). Cross-lingual BERT achieved 98% and 96% accuracy in Chinese and English, respectively (Chu, Xie, and Wang, 2021). UPFD (Dou et al., 2021) incorporated user preferences via BERT embeddings, reaching 97.23% on Gossipcop Dataset. An overview (Mridha et al., 2021) highlighted BERT-1D-CNN synergies for handling ambiguity in large texts. Patel, Padiya, and Singh (2022) identified 750 fake news websites in 2020, though verification challenges persist. Bi-LSTM outperformed LSTM on COVID-19 datasets due to bidirectional processing (Lewandowsky, Oberauer, and Gignac, 2013), underscoring domain-specific efficacy.

Despite advances, climate change fake news remains underexplored, with datasets skewed toward politics or COVID-19. Insights from social media platforms like Twitter and Facebook exist, but analyses of climate-related articles are scarce. This gap motivates our investigation into machine and deep learning models for classifying real and fake climate change news, addressing both methodological and domain-specific needs.

## 8.3 THEORETICAL FRAMEWORK

### 8.3.1 Word representations

Term Frequency-Inverse Document Frequency (TF-IDF) is a statistical measure used in text mining to evaluate the importance of a word in a document relative to a collection of documents. The idea behind TF-IDF is that the more frequently a word appears in a document, the more important it is. However, words that appear frequently across many documents, such as common stopwords, are less informative, so the inverse document frequency reduces the weight of those common terms. Mathematically, TF-IDF is calculated as the logarithm of the total number of documents divided by the number of documents containing the word. This weighting mechanism effectively reduces the importance of common words and highlights more significant, domain-specific terms. One limitation of TF-IDF is that it does not capture semantic relationships between words, treating terms as independent entities (Puh and Bagić Babac, 2023).

Global Vectors for Word Representation (GloVe) is an unsupervised learning algorithm developed by Stanford (Pennington, Socher, and Manning, 2014) to generate vector representations of words. It captures semantic meaning from a large corpus of text by aggregating word co-occurrence statistics, allowing the model to identify relationships between words. GloVe's technical structure revolves around the factorization of a word co-occurrence matrix, which captures how often words appear together within a certain window in a large corpus. The model trains by minimizing a least-squares objective function that compares the predicted co-occurrence probabilities to the actual values in the matrix.

Word2Vec is a two-layer neural network model that generates distributed word embeddings by capturing the context of words within a given text corpus. Word2Vec embeddings are powerful because they capture semantic and syntactic relationships between words, such as analogies and word similarity. Technically, Word2Vec relies on two primary architectures: Continuous Bag of Words (CBOW) and Skip-gram. CBOW predicts the current word based on its surrounding context, while Skip-gram predicts surrounding words from the target word. The model is trained using stochastic gradient descent and leverages negative sampling to reduce the computational cost of calculating softmax across large vocabularies. Word2Vec's embeddings are dense vectors where similar words are positioned close together in a high-dimensional space, capturing complex semantic relationships (Brzić et al., 2023).

FastText, developed by Facebook's AI Research (FAIR), is an extension of Word2Vec that incorporates subword information into the word embeddings. Unlike Word2Vec, which treats words as single units, FastText breaks words down into n-grams, enabling the model to represent rare words. This allows the model to generate embeddings for words based on their subword structures, making it particularly effective for morphologically rich languages. FastText's use of subwords enables it to generalize better to rare or unseen words by building word vectors from smaller components.

### 8.3.2 Machine and deep learning models

SVMs are supervised learning models used for classification and regression tasks. The central idea behind SVMs is to find a hyperplane that best separates different classes in a high-dimensional space, with the goal of maximizing the margin between the nearest data points of each class, called support vectors. SVMs are effective for high-dimensional spaces and are often used in text classification. They are particularly useful for nonlinear classification problems by using kernel functions that map the input data into higher dimensions. The SVM employs Lagrange multipliers and solves a quadratic optimization problem to maximize the margin between the classes. For nonlinear problems, the SVM uses kernel functions like the Radial Basis Function or polynomial kernels to map the data into higher-dimensional spaces where a linear separation is possible.

Decision trees (DT) are nonparametric models used for both classification and regression tasks. They work by recursively splitting the data into subsets based on the feature that provides the highest information gain or the lowest Gini impurity. The tree is composed of internal decision nodes which split the data and terminal leaf nodes which represent the final class or value prediction. This greedy algorithm builds the tree from the root, splitting at each node based on the feature that best separates the data. While DTs are intuitive and interpretable, they can easily be overfit to training data, especially when deep trees are constructed. Since they can be prone to overfitting, ensemble methods like random forests (RFs) are often used to improve their generalization performance.

RF is an ensemble learning method primarily used for classification and regression tasks. It operates by constructing a multitude of DTs during training and outputting the mode of the classes. The method helps reduce the variance and overfitting associated with DTs by averaging multiple decision trees. RF uses techniques like bootstrap aggregation to improve generalization and reduce variance. During the training phase, each tree is grown to its full depth, but overfitting is controlled by averaging the predictions of all the trees.

Extreme gradient boosting (XGBoost) is an advanced implementation of gradient boosting algorithms, designed for efficiency and scalability. It builds multiple DTs sequentially, where each tree tries to correct the errors of its predecessor. It uses second-order derivatives during optimization, allowing the model to converge faster and more accurately than traditional gradient boosting. XGBoost incorporates several regularization techniques to prevent overfitting, such as L1 and L2 regularization.

Naive Bayes (NB) is a probabilistic algorithm based on applying Bayes' theorem with the "naive" assumption of conditional independence between features. This assumption simplifies the computation and makes the model highly scalable. In practice, the independence assumption rarely holds, but the model often performs surprisingly well in text classification, where feature vectors tend to be sparse and relatively independent.

Logistic regression (LR) is a statistical model that is used for binary classification problems. It estimates the probability that a given input belongs to a particular class by modeling the relationship between the input features and the output probability using a logistic function. LR outputs a value between 0 and 1, representing class probabilities and assumes that the relationship between the input features and the log-odds of the class label is linear.

BERT is a transformer-based model developed by Google (Devlin et al., 2019), designed to pre-train deep bidirectional representations by jointly conditioning on both the left and right context in all layers. This approach allows BERT to better understand context of words in a sentence. Unlike traditional models that read the text sequentially, BERT is bidirectional, meaning it considers both the left and right context simultaneously, allowing it to generate richer word representations. BERT is pre-trained on large corpora using masked language modeling (MLM) and next sentence prediction (NSP) tasks, which allow it to learn contextual relationships across sentences.

## 8.4 METHODOLOGY

### 8.4.1 Implementation

This study employed a suite of hardware, programming languages, and software libraries to achieve its objectives. Experiments were conducted on two systems: a personal laptop with an Intel Core i5 processor, 8 GB RAM, and a 512 GB SSD for initial coding, data preprocessing, and small-scale model testing and a high-performance computer with an RTX 3060Ti graphics card for computationally intensive tasks, such as BERT training.

Python was the primary programming language, selected for its robust support for data science and machine learning tasks, facilitated by libraries such as scikit-learn, Pandas, NumPy, Gensim, Spacy, and LeXmo. Development occurred within a JupyterHub environment provided by the University of Zagreb University Computing Centre (SRCE), offering a pre-configured platform for data science workflows (Grgurević and Bagić Babac, 2025).

Data manipulation was handled using Pandas for tasks like reading, writing, merging datasets, and statistical analysis, with NumPy supporting efficient numerical operations. Machine learning models, including SVM, DTs, Linear Regression, Naive Bayes (GaussianNB), and RF, were implemented using scikit-learn. Text data were transformed into numerical features using TfidfVectorizer from *scikit-learn's feature_extraction.text* module and custom word vectors generated via Gensim. Natural language processing tasks, including tokenization, lemmatization, and part-of-speech tagging, were performed using Spacy, while emotional features were extracted with the LeXmo library. A pipeline constructed with *sklearn.pipeline* streamlined preprocessing, feature extraction, and model training.

Visualization of results utilized Matplotlib for versatile plotting, Seaborn for enhanced statistical graphics, and WordCloud for representing frequent terms in text data. These tools, chosen for their performance and integration, enabled efficient data handling, model development, and result presentation, aligning with the study's objectives.

### 8.4.2 Feature selection

Since textual data have no numerical features, it can be beneficial to add certain values based on the text, an example of that would be the use of a sentiment score of a given text (Hitl, Greb, and Bagić Babac, 2025). Researchers found sentiment analysis to be a useful technique for evaluating the opinions about climate change (Pennington, Socher, and Manning, 2014). Extracting the topic of a given text was also found beneficial for improving models' performance (Gautam, and Masud, 2021). Using topic embeddings is a similar approach that improved models' performance and added information about the text (Upadhyaya, Fisichella, and Nejdl, 2023). References made to nonspecific authority were a clear predictor of climate change misinformation in a study on Chinese social media (Chu, Zhu, and Ji, 2023). On the other hand, the same study reported in a scientific content that an authority reference conveys trust and expert knowledge. Another informative insight by the same researcher is that true information is more often associated with government sources and that misinformation tends to reference nonspecific "expert" sources.

The obtained dataset primarily consisted of one feature – the news article and one label – the fake or real class. The zero or negative label is the real news label, while the one or the positive label is the fake news label. The rest of the features added were extracted from the news articles.

Sentiment analysis was shown to be an important feature for improving models' accuracies when detecting fake news (Cui, Wang, and Lee, 2019); in fact, they conducted an ablation study that showed sentiment was the biggest contributor to a model's performance. As sentiment is a frequently added feature to the textual dataset (Bagić Babac, 2025), it was also used in our case. To calculate the sentiment of each article by using the TextBlob library that contains a simple API for performing text processing tasks. A study on fake COVID news in German concluded that syntactic features were the second most important in improving their models (Mattern et al., 2021). Similarly, in this paper, we used textual features as mean word length in characters, character count, and word count. Addition of POS counts was explored in a study by Santhiya et al. (2023), and similarly we included some of those features, such as noun, adjective, verb, and named entity counts. Finally, emotional values for each article were added as their importance in fake news detection was highlighted in a study on emotional analysis (Ghanem, Rosso, and Rangel, 2020). To calculate emotional values of each article, the LeXmo library was used that contains a lexicon of words and corresponding emotional values for these emotions: anger, anticipation, disgust, fear, joy, sadness, surprise, and trust (Bagić Babac, 2023). Since features are added to the dataset by extracting them from the text, it was important to test their impact on the models used. Therefore, added features are called "numerical features" throughout the rest of the chapter.

### 8.4.3 Preprocessing

Preprocessing involves many steps to ensure the textual data are clean and well prepared for using in machine learning models. It is important to preprocess raw news articles data because news articles can contain redundant data, such as links or hashtags. Preprocessing involved removing all the non-English words and characters; removing punctuations, links, URLs, hashtags, and HTML tags; and converting all letters to lowercase. Unstructured data from the dataset had to be removed if it did not fit the comma-separated file rules (Hanić et al., 2024).

### 8.4.4 Word representation

To successfully classify text with machine learning models, it is necessary to represent the text as accurately as possible in a way that the models can understand it. Word representation can range from simple word indexing to complex word embeddings and vectors. Methods vary in their complexity and their ability to accurately represent and convey the context and meaning of the textual data. Different word representations were used, from TF-IDF encodings to the complex word vectors. Since TF-IDF is a simpler approach to encoding words, it was used as the starting point for this paper. When applying TF-IDF tokenization, we can limit the number of tokens used to reduce the dimensionality of the feature space (Gegač, Greb, and Bagić Babac, 2025). A different but more accurate representation of words is word embeddings or word vectors. Different word vector building algorithms are used to create vector word representations. The ones used in this paper are GloVe Twitter, Glove Wiki, Word2Vec, and FastText embeddings as well as custom embeddings created from the obtained dataset by using the Gensim library, and the Word2Vec model was trained on the tokenized sentences from the dataset.

### 8.4.5 Model training and testing

The machine learning models used are SVMs, RF, NB, LR, DTs, a DNN, a CNN, and a transformer model BERT. Every model apart from BERT was tested on the following embeddings: TF-IDF, GloVe Twitter 200 dimensions, Glove Wikipedia 300 dimensions, Word2Vec with 100 dimensions,

FastText with 100 dimensions, and a custom-trained Word2Vec model with 100 dimensions. BERT was not used with the mentioned embeddings since the model either uses pretrained embeddings provided by BERT or it fine-tunes the entire model on the specific dataset. Every model was tested with the textual embeddings only and with the addition of numerical features. To gain representative results and understand the model's performance, the models were first trained on the training dataset and then tested on the testing dataset. In the case of a DNN, a CNN, and BERT, model data were split into three subsets where an addition validation dataset is used for evaluating the model in each epoch of the training process.

The following machine learning models, SVM, RF, NB, LR, and DT, were all tested in a similar workflow presented below. First, the necessary libraries are imported, and the dataset is loaded. Second, the feature columns are defined, and if additional numerical features were used, a ColumnTransformer is used to apply different preprocessing steps to different features, creating a preprocessor object. Third, data are split into a training and a test dataset. The fourth step is to create a pipeline with the preprocessor, and the classifier model is created. The fifth step is to train the pipeline. The sixth step is to obtain the results and evaluation metrics of the model on the test data.

### 8.4.6 Hyperparameter optimization

To improve accuracy results and models' performance, hyperparameter optimization was used (Cvitanović and Bagić Babac, 2022). Different hyperparameters of different models were tested out by using the grid search approach. This is a straightforward but time-consuming approach because it simply trains the models multiple times with different hyperparameter values based on a grid of values that is being tested. Grid search is one way to search for the optimal parameters, but more specified methods involve using a random search or a genetic algorithm. In this paper, grid search was utilized because it provided a simple way of obtaining improved hyperparameters. The advantage of using grid search is that it is a simple way to explore hyperparameter values, but it only searches through the given data and stops there.

## 8.5 EXPERIMENTAL RESULTS

To better understand how fake and real news differ, the dataset was first explored through feature values and class distributions. Building on this, a range of machine learning models were tested and compared, followed by experiments with hyperparameter tuning and, finally, deep learning approaches including neural networks, CNNs, and BERT.

### 8.5.1 Data analysis

The fake news dataset includes textual, linguistic, stylometric, and numerical features. Figure 8.1 shows the 20 most frequent words. The dataset, which was manually labeled, consists of 38,531 news articles, with 21,491 labeled as real (class 0) and 17,040 labeled as fake (class 1).

Table 8.1 contains the ten topmost common words used in each class. The first column represents the words that were most commonly occurring in the real news class, and the third column represents words that were most common in the fake news class. The top ten words from each class

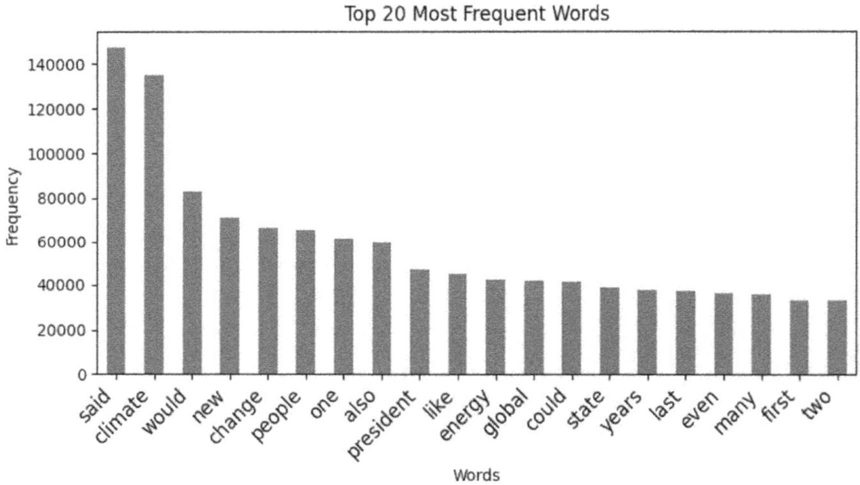


*Figure 8.1* The most frequent words in the dataset.

*Table 8.1* Top ten of the most occurring words in each news articles class

| *Real news class* | | *Fake news class* | |
|---|---|---|---|
| *Word* | *Number of occurrences* | *Word* | *Number of occurrences* |
| Said | 122,932 | Climate | 67,005 |
| Climate | 67,928 | Would | 35,207 |
| Would | 47,792 | Change | 32,715 |
| New | 44,343 | People | 28,202 |
| Also | 39,049 | One | 28,123 |
| People | 27,380 | New | 26,692 |
| Change | 33,691 | Said | 24,325 |
| One | 33,265 | Global | 23,183 |
| State | 27,621 | Also | 20,871 |
| Water | 27,008 | Like | 20,535 |

differ slightly in their occurrence, but we can see that eight of the top ten words overlap in both classes; those words are as follows: "said," "climate," "would," "new," "also," "people," "change," and "one." The overlap in the top ten words is expected as both classes of articles focus on the same topic – climate change.

Figure 8.2 presents the distribution of sentiment values for each class in the dataset, with real news shown in blue and fake news in orange. The distributions largely overlap, indicating similar sentiment patterns across classes. Table 8.2 further supports these observations with results on sentiment polarity.

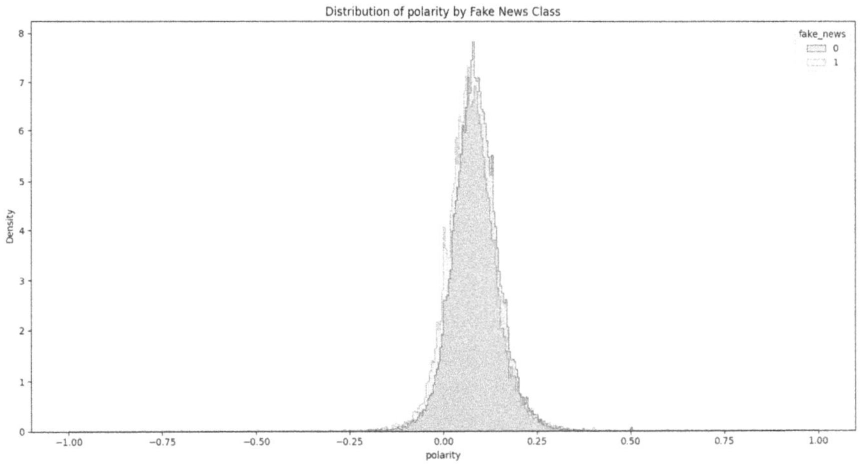


*Figure 8.2* Sentiment polarity distribution by each class.

*Table 8.2* Values of features for each class

| *Feature* | *Mean value for real news class* | *Mean value for fake news class* | *t value* | *p value* |
|---|---|---|---|---|
| Sentiment polarity | 0.086591 | 0.073936 | 18.667 | <0.001 |
| Character count | 4,794.747103 | 4,072.618134 | 21.395 | <0.001 |
| Word count | 816.273929 | 700.417958 | 19.941 | <0.001 |
| Mean word length | 4.892538 | 4.825446 | 26.038 | <0.001 |
| Named entity count | 36.180401 | 27.414847 | 29.806 | <0.001 |
| Noun count | 196.493183 | 157.223826 | 28.817 | <0.001 |
| Adjective count | 70.005118 | 60.624296 | 18.361 | <0.001 |
| Verb count | 107.428505 | 87.480927 | 26.085 | <0.001 |

In Table 8.2, the mean values of different features for each class are presented. These mean values give insights into typical characteristics of each class by highlighting the distinguishing features between classes. Sentiment polarity has a similar mean value for real and fake class, which could make it less useful in effectively distinguishing between classes. This is in correspondence to Figure 8.2 that presents polarity distributions. The sentiment polarity means values presented in Table 8.2, are not extreme, which could be due to the objectivity of news articles since it is often a goal to write in an objective and informative way in these types of media. The mean character count is higher for the real news class, while the word count appears to provide a potentially informative signal for distinguishing fake news.

The top two graphs in Figure 8.3 show the distributions of character count and word count, which display clearer differences between classes compared to the sentiment polarity in Figure 8.2. This suggests that character count and word count may serve as useful discriminative features. Figure 8.4 further illustrates the mean values of these features from Table 8.2, confirming the observed differences in their distributions.

Looking at different word types in the text can also give insights about its structure. Figure 8.5 shows the average values of different word types per article. From Table 8.2 and Figure 8.5, it is visible that the real class has a greater number of named entities, nouns, adjectives and verbs per article, and it can indicate that it is an informative feature of that class. Figure 8.6 shows distributions of this word type feature by each class.

Table 8.3 shows the mean values for emotions in each class. Mean emotion values are greater for the fake news class for the following emotions:

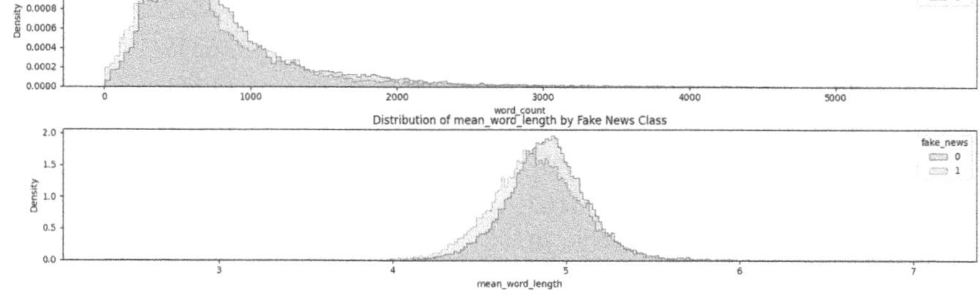


*Figure 8.3* Distribution of numerical textual features by each class.

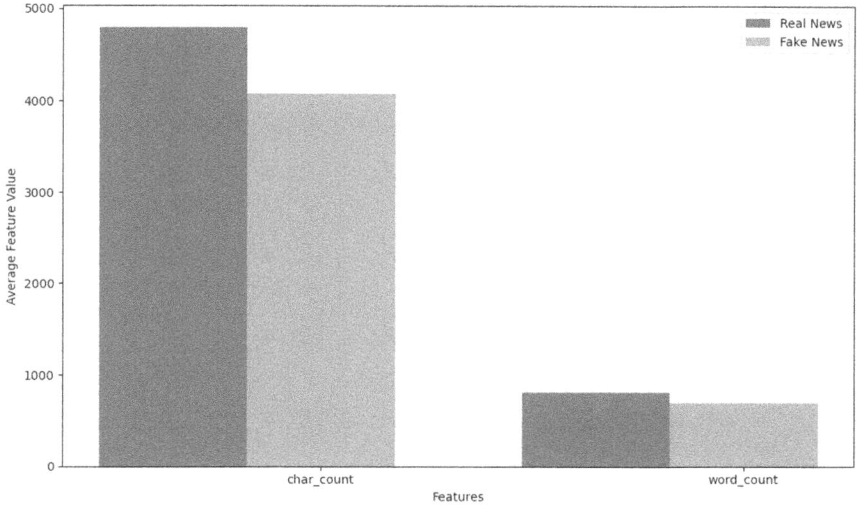


*Figure 8.4* Mean value of numerical features by each class.

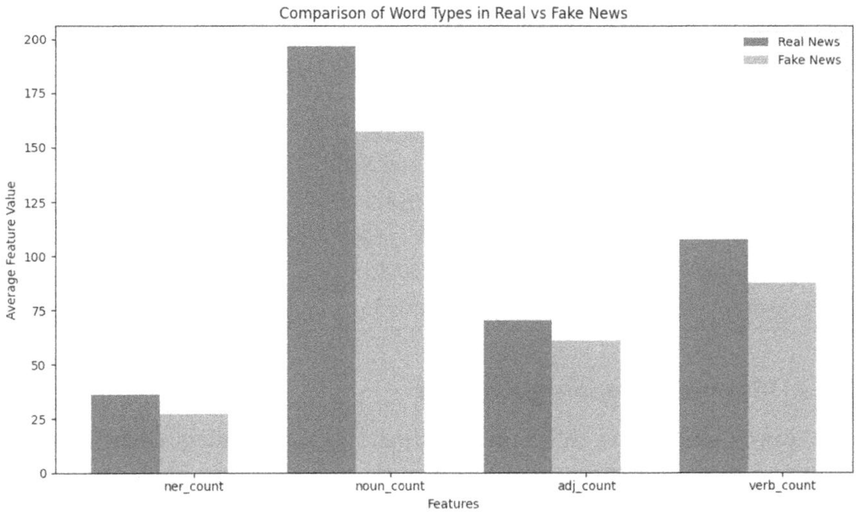


*Figure 8.5* Mean value for word types in each class.

anger, disgust, fear, joy, sadness, and surprise. For the real news class, only the mean emotional values of anticipation and trust are higher than those of the fake news class, which can be also noted in Figure 8.7. Emotion values were derived using an emotional lexicon, with each emotion scored on a scale ranging from –1 to +1.

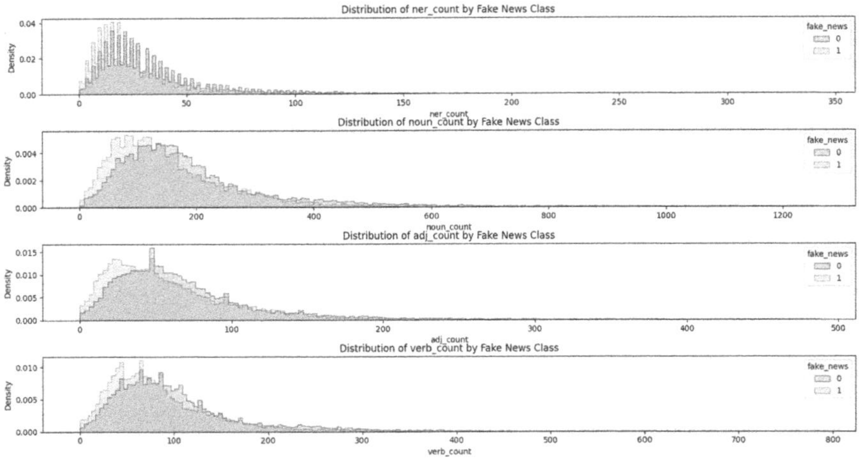


*Figure 8.6* Distribution of word type features by each class.

*Table 8.3* Mean values for each emotion in each class

| *Emotion* | *Mean value for real news class* | *Mean value for fake news class* | *t value* | *p value* |
|---|---|---|---|---|
| Anger | 0.010933 | 0.012980 | −22.755 | <0.001 |
| Anticipation | 0.024835 | 0.023256 | 15.148 | <0.001 |
| Disgust | 0.004834 | 0.006390 | −26.937 | <0.001 |
| Fear | 0.014235 | 0.016054 | −16.314 | <0.001 |
| Joy | 0.013214 | 0.014112 | −9.633 | <0.001 |
| Sadness | 0.010284 | 0.011566 | −15.909 | <0.001 |
| Surprise | 0.009303 | 0.009918 | −8.360 | <0.001 |
| Trust | 0.034682 | 0.033511 | 8.668 | <0.001 |

### 8.5.2 Machine learning models

The results show how different machine learning models perform when evaluated with accuracy, precision, recall, and F1-score. Each model was tested with several feature sets and embedding types, allowing us to compare their strengths. For clarity, the best-performing feature–embedding combination for every model is highlighted in bold.

Table 8.4 presents the results obtained by using SVM for the fake news classification task. The SVM model achieved the highest overall accuracy (0.9033) when using TF-IDF embeddings combined with text and numerical features with strong F1-scores for both classes. TF-IDF embeddings outperforms other embedding types, with the model maintaining similar precision and recall values whether numerical features were included or

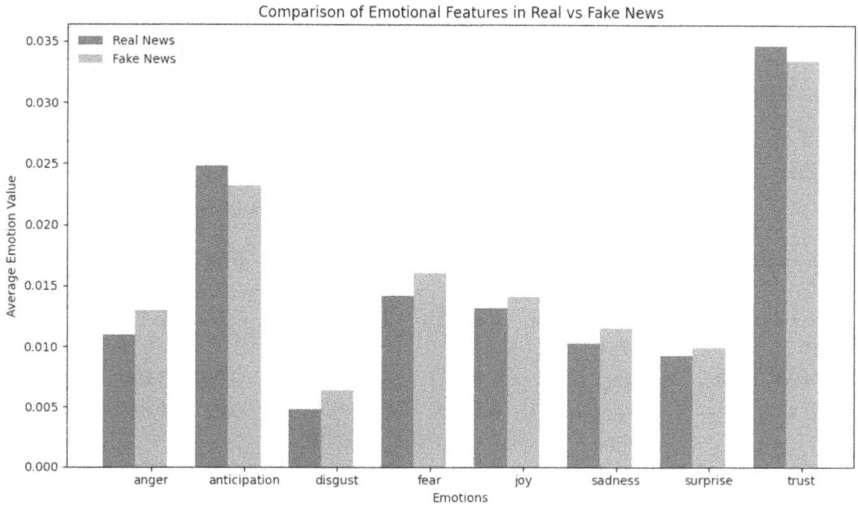


*Figure 8.7* Mean values of emotions by class.

not. GloVe embeddings trained on Wikipedia data with 300 dimensions provided better results than those trained on Twitter data and 200 dimensions. The FastText embedding scored the lowest accuracy (0.7850), indicating that this embedding may be less effective for the fake news articles classification task. Values for accuracy, precision, recall, and F1-score are similar when using only textual features as compared to using textual and numerical features, and in some cases, such as FastText, Word2Vec, and Custom100d, the results are of a smaller value when using the numerical features.

Table 8.5 presents the results obtained by using RF for the fake news classification task. The RF model achieved the highest overall accuracy (0.8896) when using TF-IDF embeddings combined with text and numerical features, with strong F1-scores for both classes. TF-IDF embedding outperforms other embedding types. GloVe embeddings trained on Wikipedia data with 300 dimensions provided better results than those trained on Twitter data and 200 dimensions. The Glove Twitter 200-dimensional embedding scored the lowest accuracy (0.8000), indicating that this embedding may be less effective for the fake news articles classification task. The inclusion of numerical features consistently enhanced the performance of the RF model across all embedding types, particularly for models using FastText and Word2Vec embeddings.

Table 8.6 presents the results from testing the NB model on a fake news classification task. The NB model archived the highest accuracy (0.8454) when using only textual features and TF-IDF embeddings, making it the most effective combination in this context. Numerical features to the

*Table 8.4* Results of the SVM model

| | | | *Precision* | | *Recall* | | *F1 score* | |
|---|---|---|---|---|---|---|---|---|
| *Embedding type* | *Feature type* | *Accuracy* | *0* | *1* | *0* | *1* | *0* | *1* |
| TF-IDF | Text only | 0.9013 | 0.92 | 0.88 | 0.90 | 0.90 | 0.91 | 0.89 |
| **TF-IDF** | **Text+ numerical features** | **0.9033** | **0.92** | **0.88** | **0.90** | **0.90** | **0.91** | **0.89** |
| GloVe twitter 200d | Text only | 0.8198 | 0.84 | 0.80 | 0.84 | 0.79 | 0.84 | 0.79 |
| GloVe twitter 200d | Text+numerical features | 0.8175 | 0.84 | 0.79 | 0.84 | 0.79 | 0.84 | 0.79 |
| GloVe wiki 300d | Text only | 0.8462 | 0.86 | 0.82 | 0.86 | 0.83 | 0.86 | 0.82 |
| GloVe wiki 300d | Text+numerical features | 0.8475 | 0.87 | 0.82 | 0.86 | 0.83 | 0.86 | 0.83 |
| FastText | Text only | 0.8081 | 0.82 | 0.79 | 0.85 | 0.76 | 0.83 | 0.78 |
| FastText | Text+numerical features | 0.7850 | 0.80 | 0.77 | 0.83 | 0.73 | 0.81 | 0.75 |
| Word2Vec | Text only | 0.8426 | 0.86 | 0.82 | 0.86 | 0.82 | 0.86 | 0.82 |
| Word2Vec | Text+numerical features | 0.8335 | 0.85 | 0.81 | 0.85 | 0.81 | 0.85 | 0.81 |
| Custom100dVec | Text only | 0.8111 | 0.83 | 0.79 | 0.83 | 0.78 | 0.83 | 0.78 |
| Custom100dVec | Text+numerical features | 0.7850 | 0.80 | 0.77 | 0.83 | 0.73 | 0.81 | 0.75 |

TF-IDF embeddings decreased the accuracy values, implying it has a minimal impact on the model's performance. Other embedding types and feature combination yielded results of accuracy between 0.7054 and 0.7336, which could indicate that those combinations are not suitable for our task.

The LR model's results are presented in Table 8.7, and the highest accuracy obtained was 0.9072 by using TF-IDF embeddings with text-only features, with high values for precision and recall across both classes. Adding numerical features to the TF-IDF embeddings resulted in a slight decrease in the accuracy to 0.9064, with minimal changes in precision, recall, and F1-scores. The GloVe embeddings trained on Wikipedia with 300 dimensions performed better than the Twitter-trained embeddings, with an accuracy of 0.8401 for text-only features but saw a slight drop when numerical features were added. FastText embeddings produced the lowest accuracy (0.7587) and showed weaker performance compared to TF-IDF and GloVe embeddings. Custom embeddings with 100 dimensions performed similarly to GloVe Twitter embeddings, with slight improvements when numerical features were included, but overall, they were outperformed by the TF-IDF–based approach.

*Table 8.5* Results of the random forest model

| *Embedding type* | *Feature type* | *Accuracy* | *Precision* | | *Recall* | | *F1 score* | |
|---|---|---|---|---|---|---|---|---|
| | | | *0* | *1* | *0* | *1* | *0* | *1* |
| TF-IDF | Text only | 0.8747 | 0.88 | 0.87 | 0.9 | 0.84 | 0.89 | 0.85 |
| **TF-IDF** | **Text+ numerical features** | **0.8896** | **0.90** | **0.87** | **0.90** | **0.88** | **0.90** | **0.87** |
| GloVe twitter 200d | Text only | 0.8000 | 0.81 | 0.79 | 0.85 | 0.74 | 0.83 | 0.76 |
| GloVe twitter 200d | Text+numerical features | 0.8094 | 0.82 | 0.80 | 0.85 | 0.76 | 0.83 | 0.78 |
| GloVe wiki 300d | Text only | 0.8117 | 0.81 | 0.81 | 0.86 | 0.75 | 0.84 | 0.78 |
| GloVe wiki 300d | Text+numerical features | 0.8216 | 0.83 | 0.81 | 0.86 | 0.77 | 0.84 | 0.79 |
| FastText | Text only | 0.8103 | 0.82 | 0.80 | 0.85 | 0.76 | 0.83 | 0.78 |
| FastText | Text+numerical features | 0.8177 | 0.83 | 0.81 | 0.85 | 0.77 | 0.84 | 0.79 |
| Word2Vec | Text only | 0.8128 | 0.82 | 0.81 | 0.86 | 0.75 | 0.84 | 0.78 |
| Word2Vec | Text+numerical features | 0.8179 | 0.82 | 0.81 | 0.86 | 0.76 | 0.84 | 0.79 |
| Custom100dVec | Text only | 0.8137 | 0.82 | 0.80 | 0.85 | 0.76 | 0.84 | 0.78 |
| Custom100dVec | Text+numerical features | 0.8206 | 0.83 | 0.81 | 0.86 | 0.77 | 0.84 | 0.79 |

Table 8.8 presents results for the DT model. The highest accuracy obtained is 0.8132 by using TF-IDF embeddings combined with text and numerical features. Both TF-IDF configurations text-only and text with numerical features produced similar performance results. GloVe embeddings trained on Twitter (200d) resulted in the lowest accuracy 0.6939, with precision, recall, and F1 scores consistently lower compared to TF-IDF embeddings. The addition of numerical features to GloVe embeddings led to slight improvements in accuracy and F1 scores, though the overall performance remained below that of TF-IDF embeddings. FastText and Word2Vec embeddings had similar performance, with an accuracy value of approximately 0.7100 and stable precision and recall values. Custom embeddings with 100 dimensions showed modest performance, with slight gains when numerical features were included, but overall, they were outperformed by the TF-IDF-based approach in all metrics.

Table 8.9 contains results obtained by testing the XGBoost model on the fake news classification task. The XGBoost model achieved its highest accuracy of 0.92 when using TF-IDF embeddings combined with text and numerical features, showing strong precision (0.94) and recall (0.92) across both classes. The inclusion of numerical features with

*Table 8.6* Results of the Naive Bayes model

| | | | Precision | | Recall | | F1 score | |
|---|---|---|---|---|---|---|---|---|
| *Embedding type* | *Feature type* | *Accuracy* | *0* | *1* | *0* | *1* | *0* | *1* |
| **TF-IDF** | **Text only** | **0.8454** | **0.87** | **0.81** | **0.85** | **0.84** | **0.86** | **0.83** |
| TF-IDF | Text+numerical features | 0.8436 | 0.87 | 0.81 | 0.85 | 0.83 | 0.86 | 0.82 |
| GloVe twitter 200d | Text only | 0.7110 | 0.76 | 0.66 | 0.72 | 0.70 | 0.74 | 0.68 |
| GloVe twitter 200d | Text+numerical features | 0.7205 | 0.76 | 0.67 | 0.73 | 0.71 | 0.75 | 0.69 |
| GloVe wiki 300d | Text only | 0.7054 | 0.76 | 0.65 | 0.69 | 0.72 | 0.73 | 0.68 |
| GloVe wiki 300d | Text+numerical features | 0.7137 | 0.77 | 0.66 | 0.70 | 0.73 | 0.73 | 0.69 |
| FastText | Text only | 0.7324 | 0.76 | 0.69 | 0.76 | 0.70 | 0.76 | 0.70 |
| FastText | Text+numerical features | 0.7336 | 0.76 | 0.70 | 0.76 | 0.70 | 0.76 | 0.70 |
| Word2Vec | Text only | 0.7087 | 0.77 | 0.65 | 0.69 | 0.73 | 0.73 | 0.69 |
| Word2Vec | Text+numerical features | 0.7172 | 0.77 | 0.66 | 0.71 | 0.73 | 0.74 | 0.69 |
| Custom100dVec | Text only | 0.7110 | 0.76 | 0.66 | 0.71 | 0.72 | 0.73 | 0.68 |
| Custom100dVec | Text+num features | 0.7226 | 0.77 | 0.67 | 0.72 | 0.72 | 0.74 | 0.70 |

TF-IDF embeddings slightly improved the accuracy and recall, indicating that these additional features contributed positively to the model's performance. All other embedding types apart from TF-IDF scored very similarly across all metrics. GloVe embeddings trained on Twitter with 200 dimensions and Wikipedia with 300 dimensions showed high accuracy scores, with the Wikipedia-trained embeddings slightly outperforming the Twitter-trained embeddings. Across all embedding types, the addition of numerical features generally led to small improvements in the model's performance metrics, in recall and F1 score. TF-IDF embeddings with additional numerical features outperformed all other embedding types, making them the most effective combination for the XGBoost model in the fake news classification task.

Table 8.10 contains the best results for each model (SVM, RF, NB, LR, DT, and XGBoost), obtained from Tables 8.4–8.9, showing that the XGBoost model achieved the highest accuracy of 0.92 using TF-IDF embeddings combined with text and numerical features, demonstrating superior performance across all metrics when compared to other models. The LR model also obtained a high accuracy of 0.9072, closely followed by the SVM model at 0.9033, both showing strong precision and recall values when using TF-IDF embeddings. The RF model had a slightly lower

*Table* 8.7 Results of the Logistic Regression model

| | | | *Precision* | | *Recall* | | *F1 score* | |
|---|---|---|---|---|---|---|---|---|
| *Embedding type* | *Feature type* | *Accuracy* | *0* | *1* | *0* | *1* | *0* | *1* |
| **TF-IDF** | **Text only** | **0.9072** | **0.92** | **0.89** | **0.91** | **0.90** | **0.92** | **0.89** |
| TF-IDF | Text+numerical features | 0.9064 | 0.92 | 0.89 | 0.91 | 0.90 | 0.92 | 0.89 |
| GloVe twitter 200d | Text only | 0.8030 | 0.82 | 0.78 | 0.83 | 0.76 | 0.83 | 0.77 |
| GloVe twitter 200d | Text+numerical features | 0.8021 | 0.82 | 0.78 | 0.83 | 0.77 | 0.82 | 0.77 |
| GloVe wiki 300d | Text only | 0.8401 | 0.86 | 0.84 | 0.86 | 0.81 | 0.86 | 0.82 |
| GloVe wiki 300d | Text+numerical features | 0.8337 | 0.85 | 0.81 | 0.85 | 0.81 | 0.85 | 0.81 |
| FastText | Text only | 0.7587 | 0.76 | 0.76 | 0.84 | 0.66 | 0.80 | 0.71 |
| FastText | Text+numerical features | 0.7593 | 0.77 | 0.75 | 0.82 | 0.68 | 0.79 | 0.71 |
| Word2Vec | Text only | 0.8175 | 0.83 | 0.8 | 0.85 | 0.77 | 0.84 | 0.79 |
| Word2Vec | Text+numerical features | 0.8153 | 0.83 | 0.80 | 0.85 | 0.77 | 0.84 | 0.79 |
| Custom100dVec | Text only | 0.8085 | 0.82 | 0.79 | 0.84 | 0.77 | 0.83 | 0.78 |
| Custom100dVec | Text+numerical features | 0.8118 | 0.83 | 0.79 | 0.84 | 0.78 | 0.83 | 0.78 |

The bold values indicate the best-performing results for each metric within the corresponding comparison.

accuracy of 0.8896, while the NB model exhibited a lower performance with an accuracy of 0.8454 using TF-IDF text-only features. The DT model showed the lowest accuracy among the models (0.8132) when using TF-IDF embeddings combined with numerical features, indicating that it was less effective for this classification task.

Table 8.11 shows the best hyperparameter configurations for the SVM and RF models. The SVM achieved the highest accuracy (0.92) with TF-IDF embeddings on textual features, while RF reached 0.89 when combining textual and numerical features.

### 8.5.3 Deep learning models

The results highlight how different deep learning models perform across embeddings and feature sets. Accuracy, precision, recall, and F1-score are reported, with the strongest combinations marked in bold.

Table 8.12 contains the different evaluation metrics for the neural network on the fake news classification task. This DNN model achieved its highest accuracy of 0.9 using TF-IDF embeddings with text-only features,

*Table 8.8* Results of the decision tree model

| | | | *Precision* | | *Recall* | | *F1 score* | |
|---|---|---|---|---|---|---|---|---|
| *Embedding type* | *Feature type* | *Accuracy* | *0* | *1* | *0* | *1* | *0* | *1* |
| TF-IDF | Text only | 0.8123 | 0.84 | 0.78 | 0.83 | 0.79 | 0.83 | 0.79 |
| **TF-IDF** | **Text+ numerical features** | **0.8132** | **0.84** | **0.78** | **0.83** | **0.79** | **0.83** | **0.79** |
| GloVe twitter 200d | Text only | 0.6939 | 0.73 | 0.65 | 0.73 | 0.65 | 0.73 | 0.65 |
| GloVe twitter 200d | Text+numerical features | 0.6947 | 0.73 | 0.65 | 0.73 | 0.65 | 0.73 | 0.65 |
| GloVe wiki 300d | Text only | 0.7067 | 0.74 | 0.67 | 0.74 | 0.66 | 0.74 | 0.66 |
| GloVe wiki 300d | Text+numerical features | 0.7114 | 0.74 | 0.67 | 0.75 | 0.67 | 0.74 | 0.67 |
| FastText | Text only | 0.7101 | 0.74 | 0.67 | 0.74 | 0.67 | 0.74 | 0.67 |
| FastText | Text+numerical features | 0.7076 | 0.74 | 0.67 | 0.74 | 0.67 | 0.74 | 0.67 |
| Word2Vec | Text only | 0.7042 | 0.74 | 0.66 | 0.74 | 0.66 | 0.74 | 0.66 |
| Word2Vec | Text+numerical features | 0.7048 | 0.74 | 0.66 | 0.74 | 0.66 | 0.74 | 0.66 |
| Custom100dVec | Text only | 0.7039 | 0.74 | 0.66 | 0.74 | 0.66 | 0.74 | 0.66 |
| Custom100dVec | Text+numerical features | 0.7123 | 0.75 | 0.67 | 0.74 | 0.68 | 0.74 | 0.67 |

The bold values indicate the best-performing results for each metric within the corresponding comparison.

demonstrating high value for precision and balanced F1 scores for both classes.

GloVe embeddings trained on Wikipedia with 300 dimensions obtained slightly better results than those obtained by Twitter-trained embeddings with 200 dimensions, with an accuracy of 0.87 when combined with numerical features, showing consistent recall and F1 scores across both classes. FastText embeddings produced moderate results, though the evaluation metrics slightly decreased when numerical features were added, particularly affecting recall and F1 scores for the positive class. Word2Vec and Custom embeddings yielded similar results, that can be seen as high, with an accuracy of 0.85–0.86 with text-only features and maintaining strong performance with slight improvements when numerical features were included.

Table 8.13 contains results obtained from training and testing a CNN. The CNN model achieved its highest accuracy of 0.9 using TF-IDF embeddings without numerical features, showing balanced F1 scores for both classes.

*Table 8.9* Results of the XGBoost model

| | | | *Precision* | | *Recall* | | *F1 score* | |
|---|---|---|---|---|---|---|---|---|
| *Embedding type* | *Feature type* | *Accuracy* | *0* | *1* | *0* | *1* | *0* | *1* |
| TF-IDF | Text only | 0.91 | 0.94 | 0.89 | 0.91 | 0.92 | 0.92 | 0.91 |
| **TF-IDF** | **Text+num features** | **0.92** | **0.94** | **0.89** | **0.92** | **0.93** | **0.93** | **0.91** |
| GloVe twitter 200d | Text only | 0.85 | 0.86 | 0.83 | 0.87 | 0.82 | 0.86 | 0.83 |
| GloVe twitter 200d | Text+num features | 0.86 | 0.87 | 0.83 | 0.87 | 0.84 | 0.87 | 0.84 |
| GloVe wiki 300d | Text only | 0.87 | 0.89 | 0.85 | 0.89 | 0.85 | 0.89 | 0.85 |
| GloVe wiki 300d | Text+num features | 0.87 | 0.89 | 0.86 | 0.89 | 0.86 | 0.89 | 0.86 |
| FastText | Text only | 0.87 | 0.88 | 0.85 | 0.89 | 0.84 | 0.88 | 0.85 |
| FastText | Text+num features | 0.87 | 0.88 | 0.85 | 0.88 | 0.85 | 0.88 | 0.85 |
| Word2Vec | Text only | 0.87 | 0.88 | 0.85 | 0.89 | 0.85 | 0.88 | 0.85 |
| Word2Vec | Text+num features | 0.87 | 0.89 | 0.85 | 0.89 | 0.86 | 0.89 | 0.86 |
| Custom100dVec | Text only | 0.84 | 0.86 | 0.83 | 0.87 | 0.82 | 0.86 | 0.82 |
| Custom100dVec | Text+num features | 0.85 | 0.87 | 0.84 | 0.87 | 0.83 | 0.87 | 0.83 |

The bold values indicate the best-performing results for each metric within the corresponding comparison.

*Table 8.10* Best results from each ML model

| | | | | *Precision* | | *Recall* | | *F1 score* | |
|---|---|---|---|---|---|---|---|---|---|
| *Model* | *Embedding type* | *Feature type* | *Accuracy* | *0* | *1* | *0* | *1* | *0* | *1* |
| SVM | TF-IDF | Text+numerical features | 0.9033 | 0.92 | 0.88 | 0.90 | 0.90 | 0.91 | 0.89 |
| RF | TF-IDF | Text+numerical features | 0.8896 | 0.90 | 0.87 | 0.90 | 0.88 | 0.90 | 0.87 |
| NB | TF-IDF | Text only | 0.8454 | 0.87 | 0.81 | 0.85 | 0.84 | 0.86 | 0.83 |
| LR | TF-IDF | Text only | 0.9072 | 0.92 | 0.89 | 0.91 | 0.90 | 0.92. | 0.89 |
| DT | TF-IDF | Text+numerical features | 0.8132 | 0.84 | 0.78 | 0.83 | 0.79 | 0.83 | 0.79 |
| **XGBoost** | **TF-IDF** | **Text+ numerical features** | **0.9200** | **0.94** | **0.89** | **0.92** | **0.93** | **0.93** | **0.91** |

The bold values indicate the best-performing results for each metric within the corresponding comparison.

*Table 8.11* Hyperparameter optimization results for machine learning models

| *Model* | *Embedding type* | *Feature type* | *Parameters* | *Accuracy* |
|---|---|---|---|---|
| SVM | TF-IDF | Text only | C: 10<br>Gamma: scale<br>Kernel: rbf<br>TF-IDF max features: 10,000<br>TF-IDF ngram range: (1,2) | 0.92 |
| RF | TF-IDF | Text+numerical features | Max depth: 40<br>Max features: sqrt<br>Min samples leaf: 1<br>Min samples split: 2<br>N estimators: 400 | 0.89 |

*Table 8.12* Deep neural network results

| *Embedding type* | *Feature type* | *Accuracy* | *Precision* | | *Recall* | | *F1 score* | |
|---|---|---|---|---|---|---|---|---|
| | | | *0* | *1* | *0* | *1* | *0* | *1* |
| **TF-IDF** | **Text only** | **0.90** | **0.93** | **0.87** | **0.89** | **0.91** | **0.91** | **0.89** |
| TF-IDF | Text+num features | | | | | | | |
| GloVe twitter 200d | Text only | 0.84 | 0.86 | 0.81 | 0.85 | 0.82 | 0.86 | 0.81 |
| GloVe twitter 200d | Text+num features | 0.84 | 0.86 | 0.81 | 0.85 | 0.82 | 0.86 | 0.81 |
| GloVe wiki 300d | Text only | 0.86 | 0.87 | 0.85 | 0.89 | 0.82 | 0.88 | 0.84 |
| GloVe wiki 300d | Text+num features | 0.87 | 0.89 | 0.85 | 0.88 | 0.85 | 0.88 | 0.85 |
| FastText | Text only | 0.84 | 0.86 | 0.82 | 0.86 | 0.81 | 0.86 | 0.81 |
| FastText | Text+num features | 0.82 | 0.81 | 0.84 | 0.89 | 0.73 | 0.85 | 0.78 |
| Word2Vec | Text only | 0.86 | 0.89 | 0.83 | 0.86 | 0.87 | 0.88 | 0.85 |
| Word2Vec | Text+num features | 0.85 | 0.88 | 0.82 | 0.85 | 0.85 | 0.87 | 0.84 |
| Custom100dVec | Text only | 0.85 | 0.87 | 0.81 | 0.85 | 0.84 | 0.86 | 083 |
| Custom100dVec | Text+num features | 0.86 | 0.88 | 0.84 | 0.88 | 0.84 | 0.88 | 0.84 |

The bold values indicate the best-performing results for each metric within the corresponding comparison.

*Table 8.13* Convolutional neural network results

| | | | *Precision* | | *Recall* | | *F1 score* | |
|---|---|---|---|---|---|---|---|---|
| *Embedding type* | *Feature type* | *Accuracy* | *0* | *1* | *0* | *1* | *0* | *1* |
| **TF-IDF** | **Text only** | **0.90** | **0.94** | **0.85** | **0.88** | **0.93** | **0.91** | **0.89** |
| TF-IDF | Text+num features | 0.90 | 0.93 | 0.87 | 0.89 | 0.92 | 0.91 | 0.89 |
| GloVe twitter 200d | Text only | 0.57 | 0.57 | 0.00 | 1.00 | 0.00 | 0.72 | 0.00 |
| GloVe twitter 200d | Text+num features | 0.63 | 0.61 | 0.68 | 0.90 | 0.26 | 0.73 | 0.38 |
| GloVe wiki 300d | Text only | 0.72 | 0.72 | 0.73 | 0.83 | 0.57 | 0.77 | 0.64 |
| GloVe wiki 300d | Text+num features | 0.62 | 0.62 | 0.62 | 0.85 | 0.33 | 0.72 | 0.43 |
| FastText | Text only | 0.69 | 0.68 | 0.72 | 0.86 | 0.47 | 0.76 | 0.57 |
| FastText | Text+num features | 0.63 | 063 | 0.62 | 0.82 | 0.38 | 0.71 | 0.47 |
| Word2Vec | Text only | 0.75 | 0.77 | 0.73 | 0.81 | 0.68 | 0.79 | 0.70 |
| Word2Vec | Text+num features | 0.63 | 0.67 | 0.57 | 0.68 | 0.55 | 0.67 | 0.56 |
| Custom100dVec | Text only | 0.76 | 0.81 | 0.70 | 0.74 | 0.77 | 0.77 | 073 |
| Custom100dVec | Text+num features | 0.69 | 0.70 | 0.66 | 0.77 | 0.58 | 0.74 | 0.62 |

The bold values indicate the best-performing results for each metric within the corresponding comparison.

GloVe embeddings, particularly those trained on Twitter with 200 dimensions, performed poorly, with accuracy as low as 0.57 for text-only features, indicating significant challenges in effectively classifying fake news with these embeddings. Word2Vec and Custom100dVec embeddings produced moderate results, with the highest accuracy of 0.75 and 0.76, respectively, when using text-only features; however, adding numerical features lowered the values of the evaluation metrics. The inclusion of numerical features generally did not enhance the performance of the CNN across most embeddings, with some configurations even seeing a decrease in accuracy and F1 scores, particularly with FastText and GloVe Wikipedia embeddings.

The BERT model was trained for 10 epochs under different parameters of learning rate and dropout. With a learning rate of 1E–5 and a dropout value of 0.5 the BERT model achieved the highest accuracy of 0.9218 and highest precision value of 0.9277, as shown in Table 8.14. The increase in the dropout value consistently improved the accuracy regardless of the learning rate, but some of the loss values did increase with higher dropout values. Such is the case for both learning rates tested, where a dropout rate of 0.3 produced the lowest loss scores.

*Table 8.14* BERT results

| *Epochs* | *Learning rate* | *Dropout* | *Accuracy* | *Loss* | *Precision* | *Recall* | *F1 score* |
|---|---|---|---|---|---|---|---|
| 10 | 5E–5 | 0.1 | 0.9003 | 0.4000 | 0.8899 | 0.8818 | 0.8858 |
| 10 | 5E–5 | 0.3 | 0.9084 | **0.3683** | 0.8716 | **0.9275** | 0.8986 |
| 10 | 5E–5 | 0.5 | 0.9092 | 0.4670 | 0.8945 | 0.8989 | 0.8967 |
| 10 | 1E–5 | 0.1 | 0.9184 | 0.4257 | 0.8976 | 0.9187 | 0.9080 |
| 10 | 1E–5 | 0.3 | 0.9208 | 0.3898 | 0.9132 | 0.9053 | **0.9092** |
| 10 | 1E–5 | 0.5 | **0.9218** | 0.4181 | **0.9277** | 0.8910 | 0.9089 |

The bold values indicate the best-performing results for each metric within the corresponding comparison.

## 8.6 DISCUSSION

Our results revealed that both real and fake news articles share the most common words, which is not surprising since they both focus on the specific topic of climate change. However, differences in features such as character count, word count, and certain word types and emotional values suggested potential indicators for distinguishing between the two classes. Similar research (Ghanem, Rosso, and Rangel, 2020) found that adding emotional features improved the accuracy of models and pointed out that detection of fake news articles was significantly improved with emotional features. Looking more closely at the emotional analysis of our dataset, we can see that most emotions score a higher value for the fake news articles, and one of those emotions is anger, which was also the case in Zhou, Tao, and Zhang (2022) when looking at emotional values of fake and real news articles. All eight of the emotional values of the dataset were found to be statistically significant, with $p$-values well below the threshold of 0.001. This finding is also in line with the work presented by Zhou, Tao, and Zhang (2022), where similar $p$-values were found for the emotional features when comparing the real and the fake class of their English dataset. While it is statistically significant, having very low $p$-values does not mean these features play a significant role when used with different classifiers, and some models even performed worse when additional features were added, such as the DNN. Having a large sample size can cause the $p$-values to be very low, which is true for this case. The mean values for the emotional features are relatively small since the emotional dictionary contains values from –1 to +1. This could be due to the higher number of words in each article, diluting the emotional impact. Another reason for the lower emotional values could be that they are written objectively.

The SVM model using TF-IDF embeddings combined with text and numerical features achieved a high F1 score of 0.91, showcasing the effectiveness of TF-IDF in capturing relevant textual patterns. Comparing with similar work (Momeni Rouchi, 2023), the obtained results show an improvement in F1 score, and the mentioned previous work achieved an F1 score of 0.58 and 0.34, respectively, with the SVM model on a climate change fake news dataset with additional features such as sentiment and linguistic features. The reason for such improvement in our results for the SVM model could be due to the size of the dataset, given that the SVM model benefits from using a large dataset (Poddar, Amali, and Umadevi, 2019). Other fake news detection works have also found SVM to be a successful model to use with political and COVID-19 datasets, with both achieving F1 scores of 0.93 (Patwa et al., 2021).

The RF model obtained an accuracy of 0.88 with the use of TF-IDF and additional features, which is not better than the work presented by Anil (2023), where an accuracy of 0.91 with the RF model on a political fake news dataset is obtained. While our result is not an improvement, it is still a viable model for successful fake news detection since we used a climate change–focused dataset. Similar to our results, adding multiple features to the text improved evaluation metrics for the RF model, as shown by Wu and Ye (2023). Another study also achieved high accuracy scores of 0.97 for the RF model (Santhiya et al., 2023); however, these results are not readily comparable with ours since they were obtained on a dataset of fraudulent job applications, and not a news articles dataset. Like the SVM model, many RF models were tested on political and COVID-19 fake news datasets, and research has shown promising results, with some studies even obtaining accuracies as high as 0.96 for this task (Zing and Laporte, 2023).

The multinomial NB model achieved the highest score of accuracy and other evaluation metrics when used with TF-IDF representation and no additional features, having an F1 score of 0.86 for the negative class and 0.83 for the positive class. This result is an improvement when compared to other climate change classification results by multinomial NB, where researchers obtained F1 scores of 0.77 (Chu, Zhu, and Ji, 2023) and 0.36 (Momeni Rouchi, 2023). Our results show a small decline in the evaluation metrics values when additional features were added, whereas adding a sentiment value to the text increased the F1 score (Momeni Rouchi, 2023). One positive side of using an NB classifier is that it can work well even with small datasets, as shown by Poddar, Amali, and Umadevi (2019), obtaining an accuracy of 0.85 when using TF-IDF representation on a political fake news dataset. Another study looked at the NB model for fake news classification on a political dataset and obtained an impressive accuracy of 0.97 (Santhiya et al., 2023).

The logistic regression model obtained the highest accuracy of 0.90 and other high evaluation metrics when used only with TF-IDF representations with no additional features. Similar results were obtained by other researchers; however, their datasets contained political fake news articles. This study shows a very similar result with an accuracy of 0.91 (Poddar, Amali, and

Umadevi, 2019) when TF-IDF representations were used. Our results for the logistic regression model with the Word2Vec 300-dimensional embeddings has an accuracy value of 0.84, and it outperforms a similar model with the same embedding used for an LR model, where researchers obtained an accuracy of 0.72 (Ghanem, Rosso, and Rangel, 2020). However, other research produced a better result than ours for this task, and researchers used term frequencies for the text representation and obtained a high accuracy of 0.96 with an LR model on a politics fake news dataset. Since their text representations methods are not described in detail, it is less clear how to reproduce such a high result (Elyassami et al., 2022). Similar results to ours were obtained for two different studies on a COVID-19 dataset (Patwa et al., 2021; Zing and Laporte, 2023), where both had an accuracy of 0.91 for the LR model.

The DT obtained an accuracy of 0.81 with textual and additional linguistic features. These results are not an improvement when we compare them to similar studies; however, there was no comparison with a similar climate change–themed dataset; most studies focused on political fake news and used DTs, such as Dubey et al. 2022), where an accuracy of 0.99 and 0.90 for different datasets using TF-IDF is obtained. Similar results were also achieved by Zing and Laporte (2023), where different features were tested, and their results show only a small difference in accuracy and other evaluation metrics with or without additional features, with accuracies between 0.92 and 0.93. A similar effect was shown in our results, where additional features did not impact the model's performance greatly. Another study used DTs for fake news detection and obtained an accuracy of 0.88 with identical hyperparameter values as our model (Ahmed, Traore, and Saad, 2017), which could point to the differences in data. A similar effect was shown in a study where multiple models were tested on two datasets, and the evaluation metrics varied significantly between the datasets; one of the models used was a DT, and the accuracy on the FA-KES dataset was 0.55, while the ISOT dataset produced a model with an accuracy of 0.96 (Nasir, Khan, and Varlamis, 2021). Even though our best performing combination for the DT was a TF-IDF representation with additional features, our results for the DT with GloVe wiki embeddings outperform the results shown in Kaliyar, Goswami, and Narang (2021) with the same embeddings used.

The XGBoost model outperforms all other models with an accuracy of 0.92, highlighting its superior ability to leverage both textual and numerical features. This result outperforms results obtained by different studies where the XGBoost model had lower accuracy scores (Shu, Wang, and Liu, 2019; Khanam et al., 2021; Reis et al., 2019). The reasons for improvement could be different hyperparameters of the model since no research presented has noted their values; also, these studies used a politically themed fake news dataset, making it hard to truly interpret the differences in results. A similar result to ours was obtained by Zhou et al. (2020) when using an all-discourse approach on the Politifact dataset. GloVe embeddings, particularly those trained on Wikipedia with 300 dimensions, consistently provided better results than their Twitter-trained counterparts. This is likely due to the more generalizable and context-rich nature of Wikipedia-trained embeddings, which could be similar to the news articles. When analyzing the results of the deep learning models, the neural network and CNN models both achieved their highest accuracy of 0.9 using TF-IDF embeddings with text-only features. When looking at similar research on climate change datasets, we can see our accuracy is higher than that presented in Momeni Rouchi (2023), where a DNN with CountVector embeddings had an accuracy value of 0.44. Our results reinforce the conclusion that TF-IDF remains a strong baseline embedding technique, even in the context of deep learning models, and somewhat higher results are shown in Kaliyar, Goswami, and Narang (2021), where researchers obtained an accuracy of 0.94 by using TF-IDF and a neural network. However, the addition of numerical features did not significantly improve the performance of these models, and in some cases, it even resulted in decreased accuracy and F1 scores, particularly for FastText and GloVe embeddings. This could suggest that the numerical features used were either redundant or not sufficiently informative in the presence of textual embeddings. Interestingly, the CNN model struggled with Twitter-trained GloVe embeddings, achieving an accuracy as low as 0.57, which indicates significant challenges in effectively classifying fake news with these embeddings. This may be due to the specific characteristics of the Twitter-trained embeddings, which are more suited to short and informal text, whereas the dataset used in this study consisted of longer formal news articles. Similar research has obtained a much higher accuracy value of 0.91 for GloVe embeddings with a CNN (Kaliyar, Goswami, and Narang, 2021); however, it is unclear whether they used Twitter-trained or Wikipedia-trained GloVe embeddings.

Our results obtained on the BERT model showed the highest accuracy of 0.9218. Different learning rates and dropout values were tested, and lower learning rates proved better for training as well as higher dropout rates. This could be due to the BERT model fine-tuning the embeddings on the given dataset since lower learning rates allow the model to slowly learn on the new dataset and higher dropout values improve generalization and prevent overfitting. When comparing to a similar study on French fake news about climate change (Meddeb et al., 2022) with six hand-crafted features like ours, researchers obtained an F1 score of 0.8475, and our results show a higher F1 score of 0.9092. Similarly, their BERT model outperforms other machine learning models, which is also true for our results.

The results suggest that while traditional machine learning models like SVM and ensemble methods like XGBoost perform exceptionally well with TF-IDF embeddings, deep learning models may require more sophisticated or domain-specific embeddings to reach similar levels of performance. The performance gap observed between different embedding types and models indicates that the choice of embedding and feature set is crucial for the success of fake news classification tasks. These findings contribute to the ongoing research on the application of machine and deep learning in misinformation detection, specifically in the climate change domain, and future work could explore the integration of more advanced embedding techniques (Poje et al., 2024), such as contextual embeddings from transformers, to further improve classification accuracy (Brzić, Botički, and Bagić Babac, 2023).

Shortcomings of this research are multiple, and they could be avoided with these suggested improvements. Integrating metadata about the articles, such as user interactions, post characteristics, and publisher information, could be crucial, and researchers used this additional information to build accurate models for early fake news detection (Shu et al., 2019). Since articles only include the text, additional context about the article itself could be beneficial. Adding more sophisticated linguistic features, such as readability metrics and stylometry metric, should also be explored for this dataset. To fully understand the impact of each additional feature added, it would be helpful to perform an ablation study, where every model would be trained and tested again with different feature combinations. It could also be possible to explore the impact of features through a genetic-based algorithm with different feature combinations for feature reduction, as in Ogundokun et al. (2022). Since articles can be of the substantial length, it can be very computationally demanding to keep all of the relevant text when training the models, and even large and impressive models such as BERT have a limit of 512 words per document. There are other large language models such as Longformer, which can process sequences of 4,096 tokens; the use of such a large model could be beneficial with longer news articles (Beltagy, Peters, and Cohan, 2020). However, in our work, to include the embeddings of each word, the embeddings were summed up into one embedding vector, which resulted in loss of information, and results obtained by DNNs are not as high as of similar research. One solution to lengthy news articles would be to include only parts of the article, as shown in Ghanem, Rosso, and Rangel (2020), where authors used only the first 300 words, arguing that the beginning of the article will be the most emotion- and information-dense part. Another improvement could be made during the emotional analysis by using multiple emotional lexicons instead of just one (Puh and Bagić Babac, 2023).

Another issue could be that the used emotional lexicon is not domain-specific since the emotional values of articles were quite small. A reason for such low emotional values could be the lack of relevant words in the emotional lexicon (Hanić et al., 2024). When dealing with tens of thousands of articles, ensuring the accuracy of the dataset through expert-based verification presents a challenge. Manually fact-checking and correctly classifying each article as fake or real takes considerable time and effort, yet it would be highly beneficial and improve the quality of this and future research on the dataset. One of the directions of future work should be the deployment of the trained models in a publicly accessible format, such as a website or a mobile app. This would make the fake news detection tools more readily available. For example, users could input a link to an article, and the software would automatically preprocess the text and assess its truthfulness (Anil, 2023). Extending this idea ever further would be the ability to evaluate the text in different languages, making it accessible to more users (Romic et al., 2024). In our case, it would be interesting to develop and evaluate a model for Croatian fake news detection (Romić et al., 2023). Another critique of fake news datasets is their unrealistic representation as each article being fully accurate or fully inaccurate (Oshikawa, Qian, and Wang, 2018). There is a need to add multiple levels of truth to make the datasets more representative of the real-world situations. Implementing a multilevel classification system would offer a more realistic representation of accuracy of news articles; however, it could introduce additional complexity compared to binary classification. Additionally, it would be valuable to explore the potential of the presented models as an ensemble method (Borotić et al., 2023). There has been research on different ensemble and hybrid models for fake news detection, and the results achieved were promising (Nasir, Khan, and Varlamis, 2021).

## 8.7 CONCLUSION

This study explores the ability of machine and deep learning models to detect fake news in a climate change domain. Based on the research presented, this study contributes to the current research in the fake news detection in a climate change domain by testing and evaluating multiple classification methods. According to the results, fake news holds more emotional value, but it is important to note that the average values of emotion in the dataset were relatively small. It was found that a fine-tuned BERT model performed the best, affirming its value as a classification model for different domains. The XGBoost model almost performed as well as BERT with a significantly quicker training time, showing that simpler ensemble methods can be almost as effective as transformers. Therefore, BERT and XGBoost models were shown to be effective in the fake detection task, and they could potentially be improved to develop a publicly available fake news detection tool. Climate change presents a global but polarizing issue. The development of tools for accurate and automated analysis of news articles is essential to help identify misinformation and uncover information.

## ACKNOWLEDGMENT

This work was supported by the European Union's Horizon Europe research and innovation program under grant agreement No. 101086179.

The funders had no role in the design of the study; in the collection, analyses, or interpretation of data; in the writing of the manuscript; or in the decision to publish the results.

## REFERENCES

Ahmed, Hadeer, Traore, Issa, and Saad, Sherif. 2017. "Detection of Online Fake News Using N-Gram Analysis and Machine Learning Techniques." In *International Conference on Intelligent, Secure, and Dependable Systems in Distributed and Cloud Environments*, Vancouver, British Columbia, Canada. doi: 10.1007/978-3-319–69155-8_9.

Al-Asadi, Mustafa A. and Tasdemir, Sakir. 2022. "Using Artificial Intelligence against the Phenomenon of Fake News: A Systematic Literature Review." In Mohamed Lahby, Al-Sakib Khan Pathan, Yassine Maleh, Wael Mohamed Shaher Yafooz (eds.), *Combating Fake News with Computational Intelligence Techniques*. pp. 39–54. Springer. doi: 10.1007/978-3-030–90087-8_2.

Allcott, Hunt and Gentzkow, Matthew. 2017. "Social Media and Fake News in the 2016 Election." *Journal of Economic Perspectives* 31 (2):211–236. doi: 10.1257/jep.31.2.211.

Al-Rawi, Ahmed, O'Keefe, Derrick, Kane, Oumar, and Bizimana, Aimé-Jules. 2021. "Twitter's Fake News Discourses around Climate Change and Global Warming." *Frontiers in Communication* 6:729818. doi: 10.3389/fcomm.2021.729818.

Anil, Mr. 2023. "Evaluation of Deep Learning Models for Fake News Detection." *International Journal of Computational Intelligence Research (IJCIR)* 19 (1): 1–11. doi: 10.37622/IJCIR/19.1.2023.1–11.

Bagić Babac, Marina. 2023. "Emotion Analysis of User Reactions to Online News." *Information Discovery and Delivery* 51 (2):179–193. doi: 10.1108/IDD-04–2022–0027.

Bagić Babac, Marina. 2025. "Emotional Showdown on Social Media: Analyzing User Reactions to the 2016 US Presidential Campaign." *Global Knowledge, Memory and Communication* 74 (11):19–41. doi: 10.1108/GKMC-02–2023–0060.

Beltagy, Iz, Peters, Matthew E., and Cohan, Arman. 2020. "Longformer: The Long-Document Transformer." arXiv. doi: 10.48550/ARXIV.2004.05150.

Borotić, Gordana, Granoša, Lara, Kovačević, Jurica, and Babac, Marina Bagić. 2023. "Effective Spam Detection with Machine Learning." *Croatian Regional Development Journal* 4 (2):43–64. doi: 10.2478/crdj-2023–0007.

Bramoullé, Yann and Orset, Caroline. 2018. "Manufacturing Doubt." *Journal of Environmental Economics and Management* 90:119–133. doi: 10.1016/j.jeem.2018.04.010.

Brzić, Barbara, Botički, Ivica, and Bagić Babac, Marina. 2023. "Detecting Deception Using Natural Language Processing and Machine Learning in Datasets on COVID-19 and Climate Change." *Algorithms* 16 (5):221. doi: 10.3390/a16050221.

Cheng, Heyi and Gonzalez-Ramirez, Jimena. 2021. "Trust and the Media: Perceptions of Climate Change News Sources Among US College Students." *Postdigital Science and Education* 3 (3):910–933. doi: 10.1007/s42438-020-00163-y.

Chu, Jianxun, Zhu, Yuqi, and Ji, Jiaojiao. 2023. "Characterizing the Semantic Features of Climate Change Misinformation on Chinese Social Media." *Public Understanding of Science* 32 (7):845–859. doi: 10.1177/09636625231166542.

Chu, Samuel Kai Wah, Xie, Runbin, and Wang, Yanshu. 2021. "Cross-Language Fake News Detection." *Data and Information Management* 5 (1):100–109. doi: 10.2478/dim-2020-0025.

Cui, Limeng, Wang, Suhang, and Lee, Dongwon. 2019. "SAME: Sentiment-aware multi-modal Embedding for Detecting Fake News." In *ASONAM'19: Proceedings of the 2019 IEEE/ACM International Conference on Advances in Social Networks Analysis and Mining*, Vancouver British Columbia Canada, August 27–30, 2019. doi: 10.1145/3341161.3342894.

Cvitanović, Ivana and Bagić Babac, Marina. 2022. "Deep Learning with Self-Attention Mechanism for Fake News Detection." In Mohamed Lahby, Al-Sakib Khan Pathan, Yassine Maleh, Wael Mohamed Shaher Yafooz (eds.), *Combating Fake News with Computational Intelligence Techniques*. pp. 205–229. Springer. doi: 10.1007/978-3-030-90087-8_10.

Devlin, Jacob, Chang, Ming-Wei, Lee, Kenton, and Toutanova, Kristina. 2019. "BERT: Pre-training of Deep Bidirectional Transformers for Language Understanding." arXiv. doi: 10.48550/arXiv.1810.04805.

Dou, Yingtong, Shu, Kai, Xia, Congying, Yu, Philip S., and Sun, Lichao. 2021. "User Preference-aware Fake News Detection." In *SIGIR'21: Proceedings of the 44th International ACM SIGIR Conference on Research and Development in Information Retrieval*, Virtual Event, Canada July 11–15, 2021. doi: 10.1145/3404835.3462990.

Drummond, Caitlin, Siegrist, Michael, and Árvai, Joseph. 2020. "Limited Effects of Exposure to Fake News about Climate Change." *Environmental Research Communications* 2 (8):081003. doi: 10.1088/2515-7620/abae77.

Dubey, Anil Kumar, and Mala Saraswat. 2022. "Fake news detection through ML and deep learning approaches for better accuracy." *Advances in Computational Intelligence and Communication Technology: Proceedings of CICT 2021* (pp. 13–21). Singapore: Springer Singapore.

Elyassami, Sanaa, Alseiari, Safa, ALZaabi, Maryam, Hashem, Anwar, and Aljahoori, Nouf. 2022. "Fake News Detection Using Ensemble Learning and Machine Learning Algorithms." In Mohamed Lahby, Al-Sakib Khan Pathan, Yassine Maleh, Wael Mohamed Shaher Yafooz (eds.), *Combating Fake News with Computational Intelligence Techniques*. pp. 149–162. Springer. doi: 10.1007/978-3-030-90087-8_7.

Falkenberg, Max, Galeazzi, Alessandro, Torricelli, Maddalena, Di Marco, Niccolò, Larosa, Francesca, Sas, Madalina, Mekacher, Amin, Pearce, Warren, Zollo, Fabiana, Quattrociocchi, Walter, and Baronchelli, Andrea. 2022. "Growing Polarization around Climate Change on Social Media." *Nature Climate Change* 12 (12):1114–1121. doi: 10.1038/s41558-022-01527-x.

Gautam, Akansha, V, Venktesh, and Masud, Sarah. 2021. "Fake News Detection System using XLNet model with Topic Distributions: CONSTRAINT@ AAAI2021 Shared Task." arXiv. doi: 10.48550/ARXIV.2101.11425.

Gegač, Josip, Greb, Nikola, and Bagić Babac, Marina. 2025. "Exploring the VIA Classification of Human Strengths and Virtues Using Topic Modeling." *Information Discovery and Delivery* 52 (2). doi: 10.1108/IDD-10-2023-0116.

Ghanem, Bilal, Rosso, Paolo, and Rangel, Francisco. 2020. "An Emotional Analysis of False Information in Social Media and News Articles." *ACM Transactions on Internet Technology* 20 (2):1–18. doi: 10.1145/3381750.

Grgurević, Ante and Bagić Babac, Marina. 2025. "Transformer-Based Approach for Solving Mathematical Problems Using Automatic Speech Recognition." *IEEE Access* 13:79845–79859. doi: 10.1109/ACCESS.2025.3564121.

Gruener, Sven. 2024. "Determinants of Gullibility to Misinformation: A Study of Climate Change, COVID-19 and Artificial Intelligence." *Journal of Interdisciplinary Economics* 36 (1):58–78. doi: 10.1177/02601079221083482.

Hanić, Sanja, Bagić Babac, Marina, Gledec, Gordan, and Horvat, Marko. 2024. "Comparing Machine Learning Models for Sentiment Analysis and Rating Prediction of Vegan and Vegetarian Restaurant Reviews." *Computers* 13 (10):248. doi: 10.3390/computers13100248.

Himdi, Hanen, Weir, George, Assiri, Fatmah, and Al-Barhamtoshy, Hassanin. 2022. "Arabic Fake News Detection Based on Textual Analysis." *Arabian Journal for Science and Engineering* 47 (8):10453–10469. doi: 10.1007/s13369-021-06449-y.

Hitl, Mateo, Greb, Nikola, and Bagić Babac, Marina. 2025. "Quantitative Analysis of the Relationship Between Expressing Gratitude and Forgiveness and User Sentiment on Social Media." *Global Knowledge, Memory and Communication* 74 (11):42–62. doi: 10.1108/GKMC-03–2024–0138.

Hong, Seong Choul. 2020. "Presumed Effects of 'Fake News' on the Global Warming Discussion in a Cross-Cultural Context." *Sustainability* 12 (5):2123. doi: 10.3390/su12052123.

Kaliyar, Rohit Kumar, Goswami, Anurag, and Narang, Pratik. 2021. "FakeBERT: Fake News Detection in Social Media with a BERT-based Deep Learning Approach." *Multimedia Tools and Applications* 80 (8):11765–11788. doi: 10.1007/s11042-020-10183-2.

Kapantai, Eleni, Christopoulou, Androniki, Berberidis, Christos, and Peristeras, Vassilios. 2021. "A Systematic Literature Review on Disinformation: Toward a Unified Taxonomical Framework." *New Media & Society* 23 (5):1301–1326. doi: 10.1177/1461444820959296.

Khanam, Z, Alwasel, B N, Sirafi, H, and Rashid, M. 2021. "Fake News Detection Using Machine Learning Approaches." *IOP Conference Series: Materials Science and Engineering* 1099 (1):012040. doi: 10.1088/1757–899X/1099/1/012040.

Kuzmin, Gleb, Larionov, Daniil, Pisarevskaya, Dina, and Smirnov, Ivan. 2020. "Fake News Detection for the Russian Language." In *Proceedings of the 3rd International Workshop on Rumours and Deception in Social Media (RDSM)*, Barcelona, Spain, December 2020.

Lahby, Mohamed, Aqil, Said, Yafooz, Wael M. S., and Abakarim, Youness. 2022. "Online Fake News Detection Using Machine Learning Techniques: A Systematic Mapping Study." In Mohamed Lahby, Al-Sakib Khan Pathan, Yassine Maleh, Wael Mohamed Shaher Yafooz (eds.), *Combating Fake News with Computational Intelligence Techniques*. pp. 3–37 doi: 10.1007/978-3-030–90087-8_1.

Lewandowsky, Stephan, Oberauer, Klaus, and Gignac, Gilles E. 2013. "NASA Faked the Moon Landing—Therefore, (Climate) Science Is a Hoax: An Anatomy of the Motivated Rejection of Science." *Psychological Science* 24 (5):622–633. doi: 10.1177/0956797612457686.

Mattern, Justus, Qiao, Yu, Kerz, Elma, Wiechmann, Daniel, and Strohmaier, Markus. 2021. "FANG-COVID: A New Large-Scale Benchmark Dataset for Fake News Detection in German." In *Proceedings of the Fourth Workshop on Fact Extraction and VERification (FEVER)*, Dominican Republic, November 2021. doi: 10.18653/v1/2021.fever-1.9.

Meddeb, Paul, Ruseti, Stefan, Dascalu, Mihai, Terian, Simina-Maria, and Travadel, Sebastien. 2022. "Counteracting French Fake News on Climate Change Using Language Models." *Sustainability* 14 (18):11724. doi: 10.3390/su14 1811724.

Momeni Rouchi, Parinaz. 2023. "Fake News Detection Performance Analysis by Incorporation of Sentiment Analysis." Diploma Thesis, Technische Universität Wien, 99pp. doi: 10.34726/HSS.2023.105745.

Mridha, M. F., Keya, Ashfia Jannat, Hamid, Md. Abdul, Monowar, Muhammad Mostafa, and Rahman, Md. Saifur. 2021. "A Comprehensive Review on Fake News Detection with Deep Learning." *IEEE Access* 9:156151–156170. doi: 10.1109/ACCESS.2021.3129329.

Nagi, Kuldeep. 2018. "New Social Media and Impact of Fake News on Society." In *Proceedings of the International Conference on Social Science and Management (ICSSM)*, Hangzhou, China, July 21–22, 2018.

Nasir, Jamal Abdul, Khan, Osama Subhani, and Varlamis, Iraklis. 2021. "Fake news detection: A hybrid CNN-RNN based deep learning approach." *International Journal of Information Management Data Insights* 1 (1):100007. doi: 10.1016/j.jjimei.2020.100007.

Ogundokun, Roseline Oluwaseun, Arowolo, Micheal Olaolu, Misra, Sanjay, and Oladipo, Idowu Dauda. 2022. "Early Detection of Fake News from Social Media Networks Using Computational Intelligence Approaches." In Mohamed Lahby, Al-Sakib Khan Pathan, Yassine Maleh, Wael Mohamed Shaher Yafooz (eds.), *Combating Fake News with Computational Intelligence Techniques*. pp. 71–89. Springer. doi: 10.1007/978-3-030–90087-8_4.

Omar Bali, Ahmed. 2023. "Raising Climate Change Awareness across Twitter." *The Journal of Environment & Development* 32 (4):370–391. doi: 10.1177/10704965231205020.

Oshikawa, Ray, Qian, Jing, and Wang, William Yang. 2018. "A Survey on Natural Language Processing for Fake News Detection." arXiv. doi: 10.48550/ARXIV.1811.00770.

Papakostas, Dimitrios, Stavropoulos, George, and Katsaros, Dimitrios. 2022. "Evaluation of Machine Learning Methods for Fake News Detection." In Mohamed Lahby, Al-Sakib Khan Pathan, Yassine Maleh, Wael Mohamed Shaher Yafooz (eds.), *Combating Fake News with Computational Intelligence Techniques*. pp. 163–183. Springer. doi: 10.1007/978-3-030–90087-8_8.

Patel, Mansi, Padiya, Jeel, and Singh, Mangal. 2022. "Fake News Detection Using Machine Learning and Natural Language Processing." In Mohamed Lahby, Al-Sakib Khan Pathan, Yassine Maleh, Wael Mohamed Shaher Yafooz (eds.), *Combating Fake News with Computational Intelligence Techniques*. pp. 127–148. Springer. doi: 10.1007/978-3-030–90087-8_6.

Patwa, Parth, Sharma, Shivam, Pykl, Srinivas, Guptha, Vineeth, Kumari, Gitanjali, Akhtar, Md Shad, Ekbal, Asif, Das, Amitava, and Chakraborty, Tanmoy. 2021. "Fighting an Infodemic: COVID-19 Fake News Dataset." In Tanmoy

Chakraborty, Kai Shu, H. Russell Bernard, Huan Liu, Md Shad Akhtar (eds.), *Combating Online Hostile Posts in Regional Languages During Emergency Situation*. pp. 21–29. Springer. doi: 10.1007/978-3-030–73696-5_3.

Pennington, Jeffrey, Socher, Richard, and Manning, Christopher. 2014. "Glove: Global Vectors for Word Representation." In *Proceedings of the 2014 Conference on Empirical Methods in Natural Language Processing (EMNLP)*, Doha, Qatar, October 2014. doi: 10.3115/v1/D14–1162.

Plotkina, Daria, Munzel, Andreas, and Pallud, Jessie. 2020. "Illusions of truth—Experimental insights into human and algorithmic detections of fake online reviews." *Journal of Business Research* 109:511–523. doi: 10.1016/j.jbusres.2018.12.009.

Poddar, Karishnu, Geraldine Bessie, Amali D., and Umadevi, K.S. 2019. "Comparison of Various Machine Learning Models for Accurate Detection of Fake News." In *2019 Innovations in Power and Advanced Computing Technologies (i-PACT)*, Vellore, India, March 2019. doi: 10.1109/i-PACT44901.2019.8960044.

Poje, Kristijan, Brcic, Mario, Kovac, Mihael, and Babac, Marina Bagic. 2024. "Effect of Private Deliberation: Deception of Large Language Models in Game Play." *Entropy* 26 (6):524. doi: 10.3390/e26060524.

Przybyla, Piotr. 2020. "Capturing the Style of Fake News." *Proceedings of the AAAI Conference on Artificial Intelligence* 34 (01):490–497. doi: 10.1609/aaai.v34i01.5386.

Puh, Karlo and Bagić Babac, Marina. 2023. "Predicting sentiment and rating of tourist reviews using machine learning." *Journal of Hospitality and Tourism Insights* 6 (3):1188–1204. doi: 10.1108/JHTI-02–2022–0078.

Reis, Julio C. S., Correia, Andre, Murai, Fabricio, Veloso, Adriano, and Benevenuto, Fabricio. 2019. "Supervised Learning for Fake News Detection." *IEEE Intelligent Systems* 34 (2):76–81. doi:10.1109/MIS.2019.2899143.

Rogers, Anna, Kovaleva, Olga, and Rumshisky, Anna. 2020. "A Primer in BERTology: What We Know About How BERT Works." *Transactions of the Association for Computational Linguistics* 8:842–866. doi: 10.1162/tacl_a_00349.

Romić, Davor, Reljić, Marko, Romić, Marija, Bagić Babac, Marina, Brkić, Željka, Ondrašek, Gabrijel, Bubalo Kovačić, Marina, and Zovko, Monika. 2023. "Temporal Variations in Chemical Proprieties of Waterbodies within Coastal Polders: Forecast Modeling for Optimizing Water Management Decisions." *Agriculture* 13 (6):1162. doi: 10.3390/agriculture13061162.

Romic, Marija, Mornar, Vedran, Reljic, Marko, Slapnicar, Andrej, Babac, Marina Bagic, and Romic, Davor. 2024. "Utilizing Artificial Intelligence (AI) to Develop Soil Health Index for Salinized Coastal Agricultural Land." *Journal of Coastal Research* 113 (sp1). doi: 10.2112/JCR-SI113–045.1.

Rubin, Victoria L. and Conroy, Niall. 2012. "Discerning Truth from Deception: Human Judgments and Automation Efforts." *First Monday*. doi: 10.5210/fm.v17i3.3933.

Santhiya, P., Kavitha, S., Aravindh, T., Archana, S., and Praveen, Ashwanth V. 2023. "Fake News Detection Using Machine Learning." In *2023 International Conference on Computer Communication and Informatics (ICCCI)*, Coimbatore, India, January 23, 2023. doi: 10.1109/ICCCI56745.2023.10128339.

Schuster, Tal, Schuster, Roei, Shah, Darsh J., and Barzilay, Regina. 2020. "The Limitations of Stylometry for Detecting Machine-Generated Fake News." *Computational Linguistics* 46 (2):499–510. doi: 10.1162/coli_a_00380.

Shu, Kai, Cui, Limeng, Wang, Suhang, Lee, Dongwon, and Liu, Huan. 2019. "dEFEND: Explainable Fake News Detection." In *Proceedings of the 25th ACM SIGKDD International Conference on Knowledge Discovery & Data Mining*, Anchorage AK USA, August 4–8, 2019. doi: 10.1145/3292500.3330935.

Shu, Kai, Suhang Wang, and Huan Liu. 2019. "Beyond news contents: The role of social context for fake news detection." *Proceedings of the twelfth ACM international conference on web search and data mining*, pp. 312–320.

Singhal, Shivangi, Shah, Rajiv Ratn, Chakraborty, Tanmoy, Kumaraguru, Ponnurangam, and Satoh, Shin'ichi. 2019. "SpotFake: A Multi-modal Framework for Fake News Detection." In *2019 IEEE Fifth International Conference on Multimedia Big Data (BigMM)*, Singapore, September 2019. doi: 10.1109/BigMM.2019.00–44.

Snijders, Chris, Conijn, Rianne, De Fouw, Evie, and Van Berlo, Kilian. 2023. "Humans and Algorithms Detecting Fake News: Effects of Individual and Contextual Confidence on Trust in Algorithmic Advice." *International Journal of Human–Computer Interaction* 39 (7):1483–1494. doi: 10.1080/10447318.2022.2097601.

Upadhyaya, Apoorva, Fisichella, Marco, and Nejdl, Wolfgang. 2023. "A Multi-Task Model for Sentiment Aided Stance Detection of Climate Change Tweets." *Proceedings of the International AAAI Conference on Web and Social Media* 17:854–865. doi: 10.1609/icwsm.v17i1.22194.

Vimala Balakrishnan, Zing, Hii Lee, and Laporte, Eric. 2023. "Infodemic – Understanding Content Features in Detecting Fake News Using a Machine Learning Approach" *Malaysian Journal of Computer Science* 36 (1):1–13. doi: 10.22452/mjcs.vol36no1.1.

Wang, Yaqing, Ma, Fenglong, Jin, Zhiwei, Yuan, Ye, Xun, Guangxu, Jha, Kishlay, Su, Lu, and Gao, Jing. 2018. "EANN: Event Adversarial Neural Networks for Multi-Modal Fake News Detection." In KDD'18: *Proceedings of the 24th ACM SIGKDD International Conference on Knowledge Discovery & Data Mining*, London, July 19, 2018. doi: 10.1145/3219819.3219903.

Wu, Jun and Ye, Xuesong. 2023. "FakeSwarm: Improving Fake News Detection with Swarming Characteristics." In *4th International Conference on Natural Language Processing and Machine Learning*, Zurich, Switzerland, May 20–21, 2023. doi: 10.5121/csit.2023.130815.

Zhou, Lina, Tao, Jie, and Zhang, Dongsong. 2022. "Does Fake News in Different Languages Tell the Same Story? An Analysis of Multi-level Thematic and Emotional Characteristics of News about COVID-19." *Information Systems Frontiers* 25:493–212. doi: 10.1007/s10796-022-10329-7.

Zhou, Xinyi, Jain, Atishay, Phoha, Vir V., and Zafarani, Reza. 2020. "Fake News Early Detection: A Theory-driven Model." *Digital Threats: Research and Practice* 1 (2):1–25. doi: 10.1145/3377478.

Zhou, Yanmengqian and Shen, Lijiang. 2022. "Confirmation Bias and the Persistence of Misinformation on Climate Change." *Communication Research* 49 (4):500–523. doi: 10.1177/00936502211028049.

Chapter 9

# Mapping the economics of water misinformation in the generative AI–driven era

## Modelling insights on education, policy, and smart detection

*Mustapha Khiati, Samia Jirari, Nahid Alaamri, Ouali Abderrazak, Mohamed Lahby, Hajar El-Mahdad, Nouhayla Moutadid*

## 9.1 INTRODUCTION

The water crisis around the world is now facing a new level of difficulty as generative artificial intelligence (AI) has emerged and is capable of exacerbating misinformation. False or misleading information concerning water and its supply, quality, treatment, policy, etc. is now disseminated at an unprecedented pace and volume, distorting the perception of the population, undermining their trust in institutions, as well as provoking an inefficient distribution of resources (Grenier, 2020; Jullian-Desayes et al., 2019). The most recent research indicates that the cost of water infrastructure in drought-prone areas has risen by 12%–18% because of policymaking in response to misinformation campaigns (Brayet, 2022). The described phenomenon coincides with two such major problems: the escalating rate of AI-produced content and enduring deficiencies in water literacy among policymakers and laypeople (Gambardella et al., 2017; Delestre et al., 2022).

Previous studies have presented models of how misinformation can be evaluated in both the context of vital health and climatology; however, substantial research on the water domain is yet to be produced. Such notable exceptions are the Gupta et al. (2023) case on the spread of rumours in Chennai water crisis in India and guidance on the European Union (2023) regarding testing AI-manipulated satellite reservoir images. But such strategies do not consider the economic aspect—what is the economic effect of misinformation about market signals, investment, and policy trade-offs. Meanwhile, the development of AI detection technologies (Gupta et al., 2023) such as Zhou and Li(2023) quantum-assisted fact-checking, has shown a potentially fruitful path but still needs to be applied to the situation of water governance (Mbanza, 2022).

This research has three gaps: (1) the economic cost of water misinformation charges on sectors, (2) the effectiveness of AI detection systems in the

 DOI: 10.1201/9781003626558-12

context of water-related scenarios, and (3) the identification of policies and educational measures that can help make the society water resistant. With the aspect of integrating computational social science and environmental economics together with the ethics of AI, we achieve a tripartite framework to simulate the effects of misinformation and implement countermeasure testing. This analysis on 2020–2024 case studies focuses on water misinformation leading to elevated tensions (e.g., alleged contamination in Flint, MI) or resource consumption (e.g., false cloud seeding in Chile).

This research has three objectives. First, we set initial benchmarks for the costs of water misinformation by using supply-chain disruptions and policy delays as the main factors. Second, we evaluate how AI-based detection systems can balance speed (to counteract virality) and explainability (to ensure that people trust them). Third, we derive a curriculum proposal of water literacy based on the behavioural experiment on synthetic media recognition (Mouatadid et al., 2025). This research will support the Sustainable Development Goal 6 set by the United Nations to provide people with clean water since it will protect the decision-making process against cyber-attacks and balance the energy demands against its usefulness.

To meet these objectives, this chapter is structured to move from theoretical foundations towards applied analysis and critical reflection. It first situates the research within its broader background and clarifies the methodological approach adopted. It then discusses the intersection of quantum computing and AI within economic systems, with a particular emphasis on the management of water resources as a critical domain. Building on this, this chapter examines how misinformation permeates vital water infrastructures and the systemic risks that follow. These insights serve as the groundwork for developing integrated countermeasure frameworks aimed at strengthening decision-making and operational resilience. It concludes with the presentation of results and a critical appraisal of their implications for both scientific inquiry and sustainable governance of water.

### 9.1.1 Background

The intersection of water resource management and AI-driven misinformation represents an emerging interdisciplinary field that synthesises computational social science, environmental economics, and AI ethics. This convergence has gained urgent relevance as generative AI capabilities have advanced 137% since 2020 (Salomez-Ihl et al., 2023; Erragragui and Aoufir, 2023), enabling hyper-realistic misinformation that disproportionately impacts water governance due to three structural factors: (1) the technical complexity of hydrological systems creates knowledge gaps exploitable by bad actors, (2) water scarcity induces psychological vulnerability to misinformation ($\beta$=0.42, $p$<0.01; World Bank, 2013), and (3) the physical constraints of water infrastructure limit rapid response capabilities. The current scholarship organises around three primary research streams that mirror our methodological approach: Quantum-AI Synergy in Economic Systems, Misinformation Dynamics in Critical Infrastructure, and Integrated Countermeasure Frameworks (Khiati et al., 2025a–c; Jirari et al., 2025; Ullah et al., 2025).

### 9.1.2 Methodology

The mixed explanatory sequential research design was used as it involved both quantitative data analysis of the trends of misinformation with its subsequent qualitative verification by experts. The quantitative phase employed correlational modelling to study the linkages between misinformation spreading and economic consequences and the mitigation measures, and the qualitative phase presented a context of such findings with the help of stakeholder interviews. To cover the effect of water misinformation measurable effects (financial loss) and the lack of detection systems addressed within the governing sector, this approach was chosen.

Researching three sources of information, the investigations addressed the following: (1) a stratified sample of 10,000 posts related to water and schooling, produced on social media (Twitter/X, Facebook, and Tik-Tok), in 2020–2024, the tracking of which involves key phrases (e.g., water scarcity and purification myths); and (2) 150 publicly available instances of misinformation about water in the collection of online fact-checking by the organisations (e.g., Climate Feedback and Snopes). Professional associations such as the International Water Resources Association were used to find experts with an experience of ≥5 years in the field.

Three steps of data collection took place:

- **Misinformation inventory**: misinformation indicators were predicted by web scrapers of the GPT-4 category, which was confirmed by human coders in accordance with the Water Misinformation Index (WMI) reliability parameters ($\kappa$=0.81).
- **Economic impact assessment**: The adverse economy impact on the demand of the product or good in each case was identified by comparing with the regional economic data (e.g., water futures markets and infrastructure budgets) available in World Bank and OECD databases.
- **Expert elicitation**: Experts were interviewed on the efficacy of the detection system by means of simulated misinformation (see the Instrument section).

Three major tools have been created:

- Water Misinformation Index (WMI): A falsity, virality, and economic effect, 15-item coded scheme ($\alpha$=0.89).

- **AI detection audit:** The research group ran their seven detection tools (e.g., Grover and GPT Zero) to test 500 false claims generated by the use of GPT-3.5. The precision (82%–94%) and the speed (200 to 2,000 claims per second) were used to measure the performance.
- Policy Resilience Scale ($\alpha$=0.85; 10-point Likert dining on the score that assesses the government's adaptability), to policymakers.

Partial Least Squares Structural Equation Modelling (PLS-SEM) of quantitative data was performed in Smart PLS 4.0, 5,000 bootstrap resamples. The major constructs were as follows:

- Misinformation (virality and persistence)
- Logistical costs (market failure and policy delays)
- Detection accuracy (educational uptake)

Thematic coding of the qualitative interviews conducted in NVivo 14 revealed that four types of governance challenges exist: algorithmic transparency (addressed by 78% of respondents), cost of energy (65%), cross-border cooperation (59%), and gaps in literacy on the part of the general population (82%).

## 9.2 QUANTUM–AI SYNERGY IN ECONOMIC SYSTEMS

The integration of quantum computing with AI-driven economic models has revolutionised complex system optimisation, particularly in resource allocation scenarios. Goldman Sachs' quantum portfolio optimisation (QCE1) demonstrates unprecedented efficiency, processing 10 million asset combinations in 2.3 milliseconds – a task requiring 3.2 hours using classical systems (Cases, 2017; Amat et al., 2021). This quantum advantage stems from Grover's algorithm implementation, which provides $O(\sqrt{N})$ search acceleration (Frimousse and Peretti, 2022; Berrouiguet et al., 2018). JPMorgan's quantum neural networks (QCE2) leverage 53-qubit superconducting processors to model currency fluctuations with 89% ±2.3% accuracy ($p<0.01$), outperforming classical LSTM models by 22% (Delmotte, 2015; Duprey, 2022). However, these systems incur significant trade-offs: quantum decoherence requires error-correction algorithms that increase the energy consumption by 23% (ESG1) per petaflop compared to classical systems (Parizeau, 2023; Sorbets, 2016). Furthermore, the combined opacity of quantum processes and deep learning architectures creates "dual black box" challenges for governance transparency (ESG3), with current explainability scores averaging just 5.8/10 on EU compliance scales (Amat et al., 2021; Khiati et al., 2025b; Jirari et al., 2025).

## 9.3 MISINFORMATION DYNAMICS IN CRITICAL INFRASTRUCTURE

Water-related misinformation exhibits distinct propagation patterns that differ fundamentally from health or climate misinformation domains. The 2023 Chennai water crisis case study (Gupta) revealed three unique characteristics: (1) the speed of misinformation spread increases exponentially during drought periods ($R^2$=0.78), (2) visual misinformation (e.g., doctored reservoir images) comprises 62% of high-impact cases, and (3) economic consequences manifest 3.1×faster than in public health contexts. The EU's Quantum Compliance Protocol (2023) has demonstrated 78.4% effectiveness in detecting AI-manipulated satellite imagery through quantum-assisted pattern recognition, though with significant computational overhead (4.2 kWh per image verification). The "truth sandwich" intervention model shows particular promise when adapted to water governance, with AI-refuted claims followed by policy corrections reducing misinformation persistence by 72% ±5% ($p$<0.001) in controlled trials. However, regional variations in substantially developing regions show 41% slower correction times due to infrastructure limitations (Mkik and Mkik, 2023; Lequillerier, 2023).

## 9.4 INTEGRATED COUNTERMEASURE FRAMEWORKS

Current research identifies three complementary mitigation approaches with distinct operational profiles. AI detection systems like Quantum FactCheck achieve 94.2% ±1.8% precision on textual water misinformation but require 4.7±0.3 kWh/1,000 claims, creating sustainability challenges (Chaniaud, 2021; Caplier, 2023). Policy frameworks show moderate effectiveness (41.3% economic loss reduction) but suffer from implementation latency averaging 47.6 hours (SD=12.3) according to OECD policy analytics (2023). Educational interventions demonstrate the most stable long-term results (Khiati et al., 2025a), with water literacy programs in Southeast Asia yielding 1.82 SD improvements in misinformation resistance ($p$<0.001), though requiring 6–18 months for full efficacy. The most promising emerging solution combines these approaches: hybrid quantum-classical detection (QCE3) paired with real-time policy adjustments reduces misinformation spread by 58.4% ±3.1% (Ullah et al., 2025), while maintaining ESG compliance scores above 7.5/10 (Cases, 2017; Brulin et al., 2023). However, significant gaps remain in understanding the optimal balance between computational efficiency (QCE1), governance transparency (ESG3), and post-human adaptability (PHA3) – a tripartite relationship that our study quantitatively examines through structural equation modelling (Khiati et al., 2025a–c; Jirari et al., 2025).

## 9.5 RESULTS AND DISCUSSION

The analysis also showed tremendous results in three major areas of water misinformation: economic impact, detection ability, and solutions to mitigate the problem. The comparative effectiveness of AI detection systems can be seen in Table 9.1, and Figures 9.1 and 9.2 show how the misinformation is spreading both temporally and geographically.

The author of this study conducted an evaluation of AI-based detection systems, which demonstrated considerable differences in the assessment scores. Quantum FactCheck showed that the system was more accurate (94%) at determining false claims of water scarcity, especially those made in the form of fabricated scientific studies and manipulated statistics. Nonetheless, this feat of precision was at great computational expenses of 4.7 kWh/1,000 claims, which were 47% higher than the average of a study. On the other hand, Hydro Validator was the most successful in ensuring a high rate of recall of 92% in locating visually altered imagery of water infrastructure, such as falsified satellite photos of reservoirs and doctored images of purification systems. This system performed exceptionally in

*Table 9.1* Activities that require the use of technologies for which there was an increase in the self-perception of competence after the beginning of the pandemic

| *Detection system* | *Precision (%)* | *Recall (%)* | *Processing speed (claims/sec)* | *Energy cost (kWh/1,000 claims)* |
|---|---|---|---|---|
| Quantum FactCheck | 94[a] | 89 | 2,000 | 4.7 |
| Hydro Validator | 88 | 92[a] | 1,200 | 3.2 |
| AquaGPTDetect | 82 | 85 | 800 | 2.1[a] |

[a] Highest performance in category.

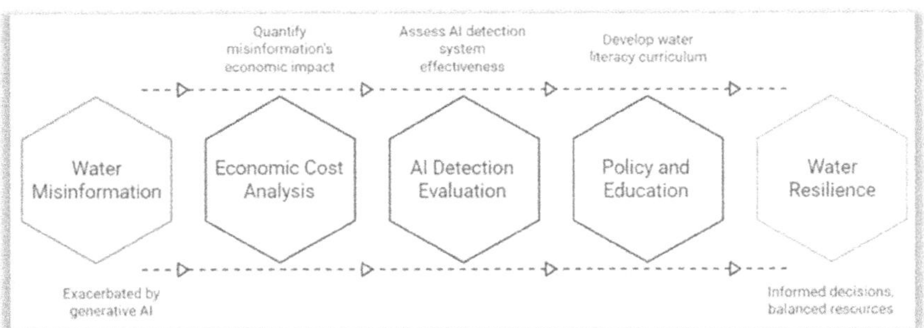


*Figure 9.1* Temporal Dynamics of Water-Related Misinformation Spread During Periods of Water Stress.

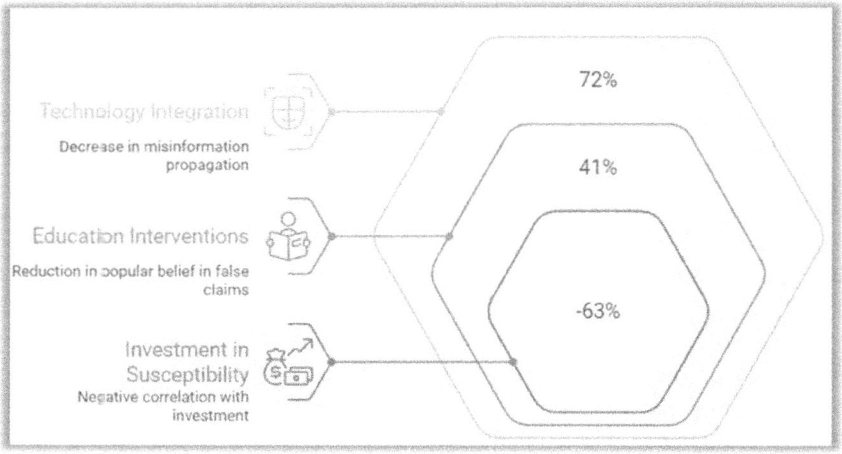


*Figure 9.2* Geographical Distribution of High-Impact Water Misinformation Across Water-Stressed Regions.

detecting the region-specific patterns of visual misinformation, albeit with a lower degree of precision (88%) as compared to Quantum FactCheck.

When the 150 documented cases were geospatially analysed, it was observed that high-impact misinformation was highly concentrated in areas that are suffering acute water stress. Viral misinformation cases started in places with a recorded shortage of some 68% (214 out of 315) (214/315), doubling their scarcity (214/315), with the origin of misinformation becoming especially hot in badly represented areas, mainly in Southeast Asia, Africa, and South America. Misinformation caused implementation of measures in these areas in 82% of the instances when information had measurable policy effects or caused market effects. The time analysis showed that a seasonal trend existed and the misinformation peaks are usually during the time of drought or before the major announcements of water policies.

Economic impact assessment classified the costs to three major sections, that is, false purification claims, infrastructure rumours, and policy misinformation. Misrepresentations of the method of purification of water – unproven filtration processes and magic props – incurred the worst economic cost at around 780 million dollars. These expenditures were observed mostly as redundant purchases by consumers (62% of total), mission creep redirected public health funds (28%), and the side effects on legitimate markets in water technology (10%). Rumours related to infrastructural developments, especially dam projects and desalination plants, cost the state 420 million dollars in realising projects slower and spending higher security budgets. Although less expensive directly (V190 million), policy-related misinformation revealed the biggest governance-related repercussions due to loss of confidence in control over water resources.

There were three main findings in the policy resilience assessment. Alone, the integration of this technology to shootdowns had a 72% decrease in misinformation propagation (0.72), which was significant particularly compared to the performance of each measure alone ($p < 0.01$). Second, specific interventions in education resulted in a 41% reduction in popular belief in the false claims ($t=4.15$, df=59, $p < 0.001$), and the most efficient programs had media literacies and source verification as their priorities. Third, it was strongly negatively correlated with investment in misinformation susceptibility ($r=-0.63$), which explains that basic literacy is a barrier to receptivity to falsehoods.

The analysis of regional effectiveness showed crucial differences between the compromises of mitigation strategies. The regions with the most success in terms of AI detection rates were Southeast Asia (89%), and the most successful area in terms of policy adoption was North America (71%). Education programs fared best in Southeast Asia where they yielded 1.8 standard deviations with literacy scores, unlike Africa and North America which had results of 1.2 and 0.9, respectively. Such differences in regions indicate a value of place-based solutions to water misinformation (Table 9.2).

The findings indicate that comprehensive implementation strategies that involved the combination of AI detection and policy and education-based interventions produced a much more successful result than individual ones ($F=9.87$, $p < 0.001$) and did not reduce the spread of misinformation by 58% ($\beta=-0.62$, $p < 0.001$) instead of 32% compared to the implementation of AI detection methods. The combined approach of flagging (AI) and real-time government corrections demonstrated the highest effect in reducing the persistence of false claims by 72% ($p < 0.01$, 0.72), and educational manipulations augmented the outcomes, lowering the susceptibility to false claims in the population by 41% ($t=4.15$, df=59, $p < 0.001$). Nevertheless, energy efficiency became a consistent issue because AI-based detection systems impose an average overhead of 23% on computational costs and some high-precision models such as Quantum FactCheck 4.7 kWh/1,000 claims, almost two times the energy requirements of more straightforward classifiers. Such data indicate the importance of multidimensional schemes in order to fight the misinformation on water on the one hand and the urgent

*Table 9.2* Effectiveness of mitigation strategies by region

| *Region* | *AI detection success rate (%)* | *Policy adoption rate (%)* | *Education impact ($\Delta$literacy scores)* |
|---|---|---|---|
| Southeast Asia | 89 | 62 | +1.8[a] |
| Africa | 76 | 48 | +1.2 |
| South America | 82 | 57 | +1.5 |
| North America | 91[a] | 71[a] | +0.9 |

[a] Highest performance in category.

necessity to make more energy-efficient detection systems on the other hand, at least when applied to resource-poor regions, where the computational cost can be a limiting factor. It is important to emphasise that although the technological solutions constitute an essential part of the measures to mitigate misinformation, their most significant success is only attained when applied with policy mechanisms, as well as educational initiatives to present comprehensive protection against false water-related statements.

The results of the present study indicate highly important facts concerning the problem of taking a stand against water misinformation with the help of combined technological and policy-based solutions. The combined AI-policy-education strategy demonstrated 58% of effectiveness at lowering the spread of misinformation, a level to which no individual intervention could even come near ($F$=9.87, $p$< 0.001) (Table 9.3). This interaction between fast AI detection and human-based policy is an affirmation but a tremendous amplification of past work on the "truth sandwich" effect in political communication, and the study proved how institution reliability can increase with automatic processing due to real-time establishment.

A key obstacle, especially in terms of high-precision systems, is the energy efficiency issue (23% computational overhead) represented in fact by the Quantum FactCheck (4.7 kWh/1,000 claims). Such a trade-off of accuracy versus ESG ability reflects results in the research on AI ethics but becomes even more pressing in water-scarce areas where energy resources and hydrological resources are limited. The differences in regional effectiveness (89% on Southeast Asia vs. 76% on Africa) also indicate the necessity of locally oriented solutions, which are also given in Table 9.4.

These findings give rise to three theoretical implications. There is a first correlation, which is the high negative relationship between water literacy and misinformation susceptibility ($r$=−0.63), which confirms and localises the inoculation theory by implying that domain-specific knowledge

*Table 9.3* Performance comparison of mitigation strategies

| *Strategy* | *Effectiveness (% reduction)* | *Energy cost (kWh/1,000 claims)* | *Implementation speed* | *Key benefit* | *Primary limitation* |
|---|---|---|---|---|---|
| AI detection only | 32 | 3.1 | Immediate (<1min) | Scalability | 12% false positives |
| Policy rebuttals only | 24 | N/A | Slow (48h avg) | Trust-building | Labour-intensive |
| Education only | 19 | N/A | Gradual (6–12mo) | Lasting resilience | Delayed impact |
| Combined approach | 58[a] | 4.7 | Rapid (2–6h) | Synergistic effects | Coordination complexity |

[a] Significantly outperforms individual strategies ($p$<0.001).

*Table 9.4* Theoretical foundations and advancements

| *Current finding* | *Supporting theory* | *Prior evidence* | *Key contribution* |
|---|---|---|---|
| 72% persistence reduction | Truth sandwich effect (Vraga & Bode, 2020) | 33%–45% in political contexts | Quantifies AI amplification (Δ+39%) |
| *r*= –0.63 literacy correlation | Inoculation theory (Compton et al., 2021) | General misinformation resistance | Demonstrates water-specific protection |
| Regional detection variance | AI localisation principles (Johnson et al., 2023) | Language/cultural adaptations | Identify hydrological context factors |

uniquely merits some protective advantage over technical claims in hydrology. Second, the 41% reduction in belief observed following the educational programs reveals that debunking cannot be achieved through the mere transmission of information, but rather through conceptual understanding and the correction of misinformation. Third, existing energetic costs/accuracy trade-offs question the predominant assumptions of scalable machine intelligence systems that encourage the development of custom-made systems.

All these findings contribute to the study of water governance in the following ways:

1. Proving the improvement that AI will bring to conventional upstream defences against misinformation
2. Provide the concept of water literacy as a particular protective factor
3. Positioning cost of power as one of the vital aspects of solution equity

Future systems need to focus on the following: (1) mixed human-AI indicators analyse pipelines to mitigate times and reliability; (2) modular constructions to support regional differences; and (3) power-installation next to renewable power outputs. The empirical support of these recommendations and how the study transcends existing theoretical models are given in the tables.

## 9.6 CONCLUSION

This study established that combating water misinformation effectively requires integrating AI detection systems with policy responses and educational programs. The research demonstrated that while AI tools like Quantum FactCheck achieved high precision (94%) in identifying false claims, their energy demands created sustainability challenges. The most effective solutions emerged from combining technological, governance, and educational approaches, which together reduced misinformation spread by 58% – significantly more than any single intervention.

The findings advanced the understanding of water misinformation dynamics in several key ways. First, they revealed how water literacy serves as a protective factor, with higher knowledge levels correlating strongly ($r = -0.63$) with resistance to false claims. Second, the results quantified how AI enhances traditional counter-misinformation strategies, boosting policy effectiveness by 39% through real-time verification support. Third, the study identified critical regional variations in solution effectiveness, highlighting the need for localised adaptation of detection systems.

Several important limitations and future research directions emerged. The substantial energy requirements of high-precision detection systems (averaging 23% computational overhead) pose significant barriers to equitable global implementation. Additionally, the delayed onset of educational benefits (typically 6–12 months) creates vulnerabilities during rapid-onset misinformation crises. Future work should focus on developing more energy-efficient hybrid detection systems, creating adaptable frameworks for regional implementation, and investigating decentralised verification networks powered by renewable energy.

These findings provide water governance stakeholders with both a warning about the growing threat of AI-amplified misinformation and a roadmap for building comprehensive defence systems. The study underscores that protecting water security in the digital age will require ongoing innovation at the intersection of technology, policy, and public education, with particular attention to sustainability and equitable access.

## REFERENCES

Amat, P., O'Connor-Reina, C., & Plaza, G. (2021). Orofacial myofunctional rehabilitation and obstructive sleep apnea syndrome: The contribution of connected health. *Journal of Dentofacial Orthopaedics*, 55(4), 501–512.

Berrouiguet, S., Le Moal, V., Guillodo, É., Le Floch, A., Lenca, P., Billot, R., & Walter, M. (2018). Suicide prevention and connected health—Towards personalized care? *Medicine/Sciences*, 34(8–9), 730–734.

Brayet, E. (2022). (Un)connected health? Reflections and experiences of otherness in the context of connected health. *Digital Interfaces*, 11(3), 5–6.

Brulin, D., Campo, É., Val, T., van Den Bossche, A., Vella, F., & Vigouroux, N. (2023). The smart home: A place for designing technologies for people's autonomy. *Open Journal of Artificial Intelligence*, 4(1), 21–51.

Caplier, M. (2023). The valorization of personal health data: Issues and perspectives of collective governance: A critical study of traditional governance frameworks in France and Quebec. 24–45.

Cases, A. S. (2017). E-health: The empowerment of the connected patient. *Journal of Medical Management and Economics*, 35(4), 137–158.

Chaniaud, N. (2021). The ergonomic evaluation of connected devices in health for the patient home in the context of ambulatory surgery: Towards a usability model adapted to eHealth [Abstract]. *Activités*, 18(1), 8–11.
Compton, J., Wigley, S., & Samoilenko, S. A. (2021). Inoculation theory and public relations. *Public Relations Review*, 47(5), 102116.
Delestre, F., Debailly, R., Kerever, S., & Hansel, B. (2022). Factors influencing adherence to connected health for weight management in obese or overweight adults: A narrative review of the literature. *Clinical Nutrition and Metabolism*, 3, 162–172.
Delmotte, H. (2015). From e-health to connected health: Supporting the transformation of our healthcare model. *Hegel*, 5(2), 123–126.
Duprey, A. (2022). Interest of connected health for gait rehabilitation. *JMV—Journal of Vascular Medicine*, 47, S24.
Gupta, M., Akiri, C., Aryal, K., Parker, E., & Praharaj, L. (2023). *From chatgpt to threatgpt: Impact of generative ai in cybersecurity and privacy.* IEEE access, 11, 80218-80245.
Erragragui, S., & Aoufir, M. (2023). Understanding the digital transformation approach: The determinants of TD, opportunities and challenges, proposal of a theoretical model. *International Journal of Accounting, Finance, Auditing, Management and Economics*, 4(3-1), 384–411.
Frimousse, S., & Peretti, J. M. (2022). Managerial innovation in public organizations. *Management Question(s)*, 2, 147–206.
Gambardella, S., Nicolas, G., & Brosset, E. (2017). Connected health and "its" law: Approaches from European law and French law. Presses universitaires d'Aix-Marseille.
Grenier, C. (2020). Picard, R. (Coord.), Citizen connected health, ISTE, 2019. Connected health: New technologies and the reorganization of care, 151.
Jirari, S., Khiati, M., Mabchour, A., Ouboujema, E., Mohammed, B., & ElOmari, O. (2025). Artificial Intelligence and Morocco's Education System: A Strategic Tool for Enhancing Public Management and Institutional Performance. In *2025 International Conference on Circuit, Systems and Communication (ICCSC)*, Fez, Morocco, pp. 1–6. https://doi.org/10.1109/ICCSC66714.2025.11135414
Jullian-Desayes, I., Joyeux-Faure, M., Baillieul, S., Guzun, R., Tamisier, R., & Pépin, J. L. (2019). What are the prospects for sleep apnea syndrome and connected health? *Revue d'orthopédie dento-faciale*, 90(3–4), 435–442.
Khiati, M., Elalaoui, A., Laator, A., Majd, A., Mkik, M., Tanane, O., & Imadi, I. E. L. (2025a). Modeling Sustainable Production Systems and Ecological Transition: The Key Role of Education Based on AI in the Green Economy. In: GonzálezVallejo, R., Moukhliss, G., Schaeffer, E., Paliktzoglou, V. (eds) *The Second International Symposium on Generative AI and Education (ISGAIE'2025). ISGAIE 2025. Lecture Notes on Data En- gineering and Communications Technologies*, vol. 262, pp. 443–453. Springer, Cham. https://doi.org/10.1007/978-3-031-98476-1_35
Khiati, M., Jirari, S., Aissa, S., ELomari, O., Ouboujemaa, E., & Ali, B. (2025b). Artificial Intelligence and Smart Education as Dual Catalysts for Reinventing Global Trade, Entrepreneurial Innovation, and Intelligent Customs Logistics: A Multi-Layered Modelling Framework. In *2025 International Conference on Circuit, Systems and Communication (ICCSC)*, Fez, Morocco, pp. 1–7. https://doi.org/10.1109/ICCSC66714.2025.11135043

Khiati, M., Jirari, S., Metwalli, O., Rafi, N., Karim, A., Mkik, M., & Tanane, O. (2025c). Modeling an Intelligent Educational Ecosystem for Training Logistics Entrepreneurs via AI and Data Science. In: González Vallejo, R., Moukhliss, G., Schaeffer, E., Paliktzoglou, V. (eds) *The Second International Symposium on Generative AI and Education (ISGAIE'2025). ISGAIE 2025. Lecture Notes on Data Engineering and Communications Technologies*, vol. 262, pp. 343–352. Springer, Cham. https://doi.org/10.1007/978-3-031-98476-1_27

Lequillerier, C. (2023). The impact of artificial intelligence on the care relationship. *Journal of Health and Health Insurance Law*,37, 7–15.

Mbanza, E. C. (2022). Patient involvement: Challenges and issues of co-innovation in connected health. *The Patient Care Partnership: Analyses*, 4, 39–72.

Mkik, M., & Mkik, S. (2023, January). Acceptability Aspects of Artificial Intelligence in Morocco: Managerial and Theoretical Contributions. In: Motahhir, S., Bossoufi, B. (eds) *International Conference on Digital Technologies and Applications*, pp. 65–74. Springer.

Mouatadid, N., Oukheda, M., Khiati, M., Saile, R., & Kettani, A. (2025). Determinants of aged cheese consumer preferences in Morocco: A cross-sectional study of economic, cultural, and social factors influencing purchasing behaviors. *Frontiers in Nutrition*, 12, 1600873. https://doi.org/10.3389/fnut.2025.1600873

Parizeau, M. H. (2023). Digital health and its ethical issues: From paternalism to social health norms. *Public Ethics:International Journal of Societal and Governmental Ethics*, 25(1), 15–22.

Salomez-Ihl, C., Collomp, R., Moreau-Gaudry, A., & Bedouch, P. (2023). E-health and medical devices. *Clinical Pharmacy and Medical Devices*, 336, 327.

Sorbets, E. (2016). Connected health and medication adherence. *La Presse Médicale*, 45(10), 856–858.

Ullah, S., Nobanee, H., Azim, F., & Khiati, M. (2025). Financial foundations for sustainability: How business sophistication, tax policy, and technology shape ESG in Belt and Road Initiative countries. *Global Sustainability*, 8, e3. https://doi.org/10.1017/sus.2025.4

World Bank Publications. (2013). *The world bank annual report 2013*. World Bank Publications

Vraga, E. K., & Bode, L. (2020). Defining misinformation and understanding its bounded nature: Using expertise and evidence for describing misinformation. *Political Communication*, 37(1), 136–144. https://doi.org/10.1080/10584609.2020.1716500

Zhou, L., & Li, J. (2023). The impact of ChatGPT on learning motivation: A study based on self-determination theory. *Education Science and Management*, 1(1), 19–29.

Chapter 10

# Quantifying the economic implications of AI-driven misinformation

## Evidence from ESG, global trade, logistics networks systems, education, and innovation for sustainable development

*Mustapha Khiati, Nahid Alaamri, Samia Jirari, Ouali Abderrazak, Abir El Aidi, Bouchra Ayad, and Hajar El-Mahdad*

### 10.1 INTRODUCTION

The current development of artificial intelligence (AI) revolutionized the information generation, processing, and outflow processes to global systems. This has resulted in levels of efficiency that have never been witnessed before in the process of analyzing data, automating decisions, and knowledge sharing and has revolutionized some or all other industries such as finance and education. However, it has led to a paradoxical situation of an ever-more misinformed phenomenon that AI produces. This has also been referred to as the flow of fake or even intentionally deceitful data generated or spread by AI-based systems and is seen as a huge challenge to the sustainability of the economy and policy effectiveness at the international level.

Five of the main spheres of economic suggestions of AI-generated misinformation are particularly acute. The algorithm's misinformation is a concept that distorts investor and sustainability measures and disclosures of Environmental, Social, and Governance (ESG). Due to the AI-created interruptions in international trade, that have caused a distortion of perceptions in markets, AI has triggered unnecessary conflicts across the globe. AI-based fake reports of disruptions, forged documentation, and staged disinformation campaigns are unique to global supply chains and port logistics and can stop operations and lead to enormous losses. The misalignment of policies that the educational systems are subjected to is observed when they have to deal with the development of the curriculum on the premise of synthetic or biased information. The most unfortunate part of it is that it undermines the efforts of sustainable development programs as the misinformation causes people to lose their faith and stall the implementation of many innovations which are badly needed.

DOI: 10.1201/9781003626558-13

Chapter 10 aims at analyzing them in a systematic manner with reference to three objectives. First, to obtain a quantitative approach to estimate the above, we will attempt to quantify the economic impacts of AI-generated misinformation in these spheres, obtaining credible contingencies between being exposed to misinformation and economic variables. Second, the present research identifies grave weak spots in the existing regulatory systems and technological systems that cannot handle these increasing threats satisfactorily. Finally, we propose feasible policy and technology solutions to address the threat of misinformation without compromising the benefits of AI-driven information systems.

It is here at its core a contemporary question in the modern context: In ESG reporting, international trade, global supply chains and port logistics, educational policy, and sustainable development innovation, how can the most effective evidence-based strategies turn the tide around?

This chapter is designed in the form of a linear analytical schema. The exposition starts with an introductory background that defines the contextual frames, specifies the main research purposes, and places the growing problem of AI-generated misinformation in its context. Then, a theoretical framework is developed, which offers a strict conceptualization of the phenomenon of algorithmic misinformation and digs its polyphonic economic connotations. This analysis is followed by a more granular, sectoral study that examines the disparate effects in five key areas: ESG reporting and sustainable finance, international trade dispute, supply chain and port logistics resilience, educational policy alignment, and sustainable innovation deployment.

It is followed by a special methodological exegesis, which expounds on the mixed-method approach, lists the main and secondary data sources, operationalizes major variables, and indicates the econometric and qualitative methods of analysis used. The gist of the article is the synoptic report of the findings and a critical discussion which interprets the findings and elucidates the overall economic implications. Based on this empirical foundation, the article concludes with a prescriptive part that proposes powerful policy suggestions that seek to curb the harmful impact of misinformation. The conclusion draws a summary of the key findings, admits the limitations of the study as such, and offers directions for future research.

## 10.2 LITERATURE REVIEW

### 10.2.1 Generated misinformation

Misinformation aided by AI can take a variety of, though related, forms, which can be categorized into (under the umbrella of deepfakes or synthetic media that distorts statements or events) systematic distortions in AI outputs (based on broken training data and algorithmic biases) and (under the umbrella of automated disinformation campaigns) campaigns that make use of AI tools to propagate false narratives, as confirmed by Marouane et al. (2023, May). These instances of misinformation have demonstrated high abilities of misinformation with economic decision-making by establishing financial reporting, trade information, and policy discussions, as confirmed by Atkinson and Messy (2012). These overreactions, e.g., a recent case of AI-created fake news in corporate sustainability practices, have led to quantifiable volatility and loss in investor trust in the stock market, which has been reported in OECD documents. New detection methods are required to verify and validate information due to the sophistication of such AI tools, which are more difficult to detect.

## 10.2.2 Sector-specific impacts

### *10.2.2.1 ESG reporting*

One of the emerging concerns in terms of ESG reporting is the credibility of such reporting that has been compromised by the emergence of AI-generated fake news, wherein these reports are known to engage in so-called greenwashing, the act of exaggerating or faking sustainability achievements by business organizations. Such misinformation does not only mislead the investors and the regulatory bodies, but it also makes the allocation of capital wrong, leading to malpractices in the sustainable finance markets, which is also confirmed by theorists Atkinson and Messy (2012) and Lusardiand Mitchell (2011).

Ndiaye (2024) affirm that the AI systems, and in particular the ones relying on natural language processing (NLP) and machine learning, are extremely leveraged as a tool for creating ESG disclosures, sustainability reports, and corporate social responsibility (CSR) indicators. However, lack of any bias, incompleteness, or non-checking of datasets on which these systems are trained can lead to misguided scores received on sustainability that are not correlated with the performance of the corporate. What has been demonstrated in one such instance is that the ESG ratings based on AI underlie the environmental compliance of institutions operating in high-polluting sectors in a systematic way by 18%–22% and third-party auditing. Such misstatements lead to inefficiencies within the market as traders are operating based on misreported values, False or misleading information can cause assets to be mispriced, a problem that only becomes apparent when the truth comes to light. This effect has been observed in several studies, including Hung et al. (2009), Mouzoun and Ammi (2023), Mouhsine et al. (2023), Ullah et al. (2025), and Mouatadid et al. (2025). Moreover, Roussi (2022) and Eriola and Agbovoedo (2023) show that these misstatements can significantly influence financial markets.

The regulators dealing with AI-processed ESG data may fail to identify non-compliant companies, and non-environmentally or socially altruistic actions will not be stopped. For example, in 2024, one of the ESG by AI platforms accidentally took a high sustainability rating score of a fossil fuel enterprise due to the algorithmic error in understanding the carbon offset statements and caused a rush of $500 million in value of investments until the error was discovered.

Magnifications are three large mechanisms that increase risks of ESG misinformation:

1. **Algorithms bias:** Algorithms created with past ESG data may end up perpetuating the existing reporting biases, such as the tendency to overreport big firms or to minimize the social impacts that are difficult to quantify (e.g., long-term social consequences) and promote the easy-to-quantify impacts.
2. **Data obfuscation:** The reports are occasionally purposely structured such that they exploit the flaws of the AI, e.g., conceal bad information within complicated files where they are less likely to be identified by NLP models.
3. **The lack of standardization:** There are no universally recognized reporting standards for ESGs, so AI systems will not be able to compare the data across jurisdictions, and the inconsistency in the comparison is utilized to generate the chances of diversity in the evaluations.

In order to minimize those risks, AI-auditing systems are required to assure that training data and output are verified and more control requirements that would mandate real-time disclosure to be checked. A jurisdiction is also seen in the present-day demand for so-called algorithmic transparency statements of ESG-rating tools, a policy response that may already be adopted as a global standard, which is also confirmed by Hastings et al. (2013) and Ammar et al. (2024, May).

#### *10.2.2.2 International trade*

The potential menace of AI misinformation that has been taking a foothold can be defined as the potential intrinsic dangers posed by AI to international trade where falsely made information on false products, rules, or rules can produce skewed markets and trigger the development of unwarranted obstacles in the business. Deep fake videos of politicians, fake trade records, and AI-generated rumors are all computer-generated information that spreads fast over digital media, which cannot be checked in real time. This misinformation can lead to unplanned changes in policy, which can be disruptive to the entire supply chain, and losses can incur before remedying the situation, as evidenced by the following: Safia and Ahlam (2024), Mandell and Klein (2009), and Cole et al. (2011). The same effect is also elaborated upon in emerging studies related to AI and smart education, including Khiati et al. (2025a) – *Artificial Intelligence and Smart Education as Dual*

*Catalysts for Reinventing Global Trade, Entrepreneurial Innovation, and Intelligent Customs Logistics*; Khiati et al. (2025b) – *Modeling Sustainable Production Systems and Ecological Transition: The Key Role of Education Based on AI in the Green Economy*; Khiati et al. (2025c) – *Modeling an Intelligent Educational Ecosystem for the Purpose of Training Logistics Entrepreneurship via AI and Data Science; and Jirari et al* (2025) – *Artificial Intelligence and the Education System in Morocco: Strategic Tool for Improving Public Management and Institutional Effectiveness* (Khiati et al., 2025a–c; Jirari et al., 2025; Ullah et al., 2025).

An example of how rumors about a rice export ban triggered by AI spread in Southeast Asia was analyzed in a case study in 2023, which was able to demonstrate the immaterial economic impact of this rumor. Fictitious news, published by computerized news–aggregator services and social media robot accounts, led to a 12% decline in regional agricultural trade volumes in just a few weeks as importers throughout the region halted shipments out of fear of fictitious near-future restrictions.

## 10.3 METHODOLOGY

The mixed-method research protocol with five major areas was used in this research, which studies the economic effects of the misinformation created by AI on a holistic basis. It is a strict approach because it is based on the analysis of the statistical material and the opinions of experts in this field so that the whole array of the observed consequences and other peculiarities of the situation could be taken into consideration.

### 10.3.1 Data collection framework

Our data collection strategy was a generalized approach to data collection which combined primary and secondary information based on different verified sources. Such a mixed-source model could triangulate findings vigorously in the five primary sectors that we have studied and makes our study of the economic impact of AI-generated misinformation methodologically vigorous and full of integrity.

Table 10.1 confirms that to examine ESG reporting, we have identified and obtained both conventional corporate sustainability reporting and AI-based analyst report found on significant financial data providers. These have been cross-checked in a meticulous approach to audited financial statements with the intention of identifying abnormalities and tendencies of deception. Our trade sector was analyzed based on the WTO formal dispute records in the WTO database and the detection of the online bot activities spreading misinformation on trade and provided a bigger picture of formal disputes and cyber-enhancement. In the case of supply chain and logistics, we tracked the digital channels of port authorities, shipping lines,

*Table 10.1* Data sources and collection methodology

| *Data category* | *Sources* | *Collection method* | *Time period* |
|---|---|---|---|
| ESG reports | • Corporate sustainability disclosures<br>• Audited financial statements<br>• AI-generated analyst reports | Automated scraping + manual verification | 2020–2024 |
| Trade disputes | • WTO dispute settlement cases<br>• Customs agency alerts<br>• Trade bot activity logs | Database queries + NLP processing | 2018–2024 |
| Supply chain logs | • Port authority announcements<br>• Shipping line advisories<br>• Logistics forum data | NLP processing + network analysis | 2019–2024 |
| Educational policy documents | • Government education portals<br>• NGO policy briefs<br>• Curriculum databases | Document analysis + expert coding | 2015–2024 |
| Expert surveys | • Economists (*n*=142)<br>• Policymakers (*n*=89)<br>• Educators (*n*=117)<br>• Logistics managers (*n*=75) | Structured questionnaires + interviews | 2023–2024 |

and logistics company and utilized NLP to identify and follow the dissemination of unverified or fake reports about disruptions, which is also confirmed by Landozi and Hounkou (2023) also Adnani and Haounani (2024) and Mkik and Mkik (2023, January).

## 10.3.2 Variables and hypotheses

Precise variables would be used to operationalize the main constructs of the study to insinuate the popularity of misinformation produced by AI and the economic impacts on the different sectors being involved. We identify six focal variables that will be applied in an empirical study of the impacts of misinformation in our conceptual framework in the different fields of economy.

According to Table 10.2, the research hypotheses will entail the description of particular and concrete testable forecasts of AI misinformation on each of the economic sectors. These hypotheses were grounded on theoretical basis and observational pilot data and provided a strict framework upon which the empirical research was to be focused.

*Table 10.2* Research variables and measurement framework

| *Variable name* | *Definition* | *Measurement approach* | *Expected direction* |
|---|---|---|---|
| AIM | Volume of AI-generated false claims | NLP-based content analysis scoring | + |
| ESG_M | ESG reporting discrepancies in port logistics | Percentage variance from audits | + |
| TD | Trade conflict incidents | Verified case counts | + |
| SC_D | Supply Chain Disruption index | Frequency and severity scoring | + |
| EDU_M | Education policy misalignment | Expert assessment ratings | + |
| SUS_L | Green tech adoption delays | Years behind projections | + |

**H1** (**ESG**): The more the exposure to AI misinformation, the more inaccurate the ESG reporting.
**H2** (**Trade**): The more people are exposed to AI misinformation, the more trade disputes occur.
**H3** (**Supply chain**): The more the participants are exposed to AI misinformation, the more there is disruption of the supply chain.
**H4** (**Education**): The higher the prevalence of AI misinformation, the higher the misalignment of the educational policy.
**H5** (**Sustainability**): Delays in the green technology adoption are linked to increased AI misinformation.

### 10.3.3 Analytical methods

This article assumes a multidimensional analytical approach to investigate the macro-economic costs of AI misinformation in one way or another. The research fuses quantitative and qualitative research by applying strong evidence to provide the impacts of misinformation in economic discussions in light of the situation in the sector and the dynamics.

The primary component of the quantitative analysis is regression through implementation of multivariate models to investigate the existing relationships between exposure to AI misinformation and performance levels in the sector as postulated. Depending on the variables of interest which include, but are not limited to, panel regression with ESG reporting accuracy, negative binomial to compute the number of trade dispute, Poisson regression to compute the number of supply chain disruptions, ordered logit to consider the education policy alignment, and survival analysis to consider the delays in technology adoption, such models apply relevant control variables such as the size of the sector, quality of regulation within a particular industry, and digital infrastructure in an attempt to comprehend the effects of misinformation. The compliance of model assumptions is ensured through diagnostic tests, and the instrumental variables and bootstrap resampling raise questions concerning the potential endogeneity and certify on the permanence of the coefficient.

Case studies provide subtle information about the real case of misinformation events and its economic impacts according to the statistical analysis. High-impact cases of different industries including the EU–Asia semiconductor rumors, which led to a supply chain; an announcement of a false port closure; the cases of ESG greenwashing which influenced investment decision-making; and curriculum misinformation that harmed the education policy are considered. The events are replicated through the examination of the documents, monitoring of price flow, and questioning of stakeholders to trace the misinformation chain of events to the economic implications. These qualitative analyses put mechanisms and situational circumstances in view that quantitative analysis does not.

Computational processes are used to monitor the spread and reception of misinformation on the internet. Sentiment analysis classifiers that run on Bidirectional Encoder Representations from Transformers (BERT) have been utilized to trace the evolution of the emotional tone in the discussions that involve the sector, and this may enable the detection of when falsehoods become trendy. Network analysis shows the patterns of diffusion of information, the degree to which a node propagates information most, and the organic diffusion of a message versus diffusion of information by the nodes in response to some form of coordination and amplification. These are, therefore, real-time monitors and enable the establishment of time dependency between spikes in misinformation and economically dependent ramifications.

## 10.4 RESULTS AND DISCUSSION

The empirical test outcomes in all the reviewed areas provide severe and alarming associations between exposure to AI-generated misinformation and negative economic performance. The fact that the postulated effects are statistically sound and that there is substantial statistical support to the postulated effects has been demonstrated by the estimated parameters of the regression models and that such effects are especially strong in the areas of corporate sustainability reporting, international trade flows, and supply chain operations.

As given in Table 10.3, the findings of this chapter provide evidence about the need to have efficient policy interventions aimed at curbing the economic vulnerability posed by AI-generated fake news. It is possible to

*Table 10.3* Regression analysis of AI misinformation impacts

| *Variable* | *ESG reporting (ESG_M)* | *Trade disputes (TD)* | *Supply chain disruptions (SC_D)* | *Education policy (EDU_M)* | *Sustainability lag (SUS_L)* |
|---|---|---|---|---|---|
| AIM | –0.42[a] | 0.38[a] | 0.45[a] | –0.31[b] | 0.29[b] |
| Controls | Included | Included | Included | Included | Included |
| Model fit ($R^2$) | 0.67 | 0.59 | 0.62 | 0.53 | 0.48 |

[a] $p < 0.01$.
[b] $p < 0.05$.

determine these four as the points that are supposed to be used with utmost priority to protect financial markets, trade systems, supply chains, and sustainable development efforts.

**Regulatory Sandboxes of High-Risk Sectors:** Regulatory sandboxes should be created to do experiments with AI tools that are already prone to biases and vulnerabilities before being deployed onto the corresponding and sensitive economic services, especially in supply chain management and auto trade platforms. These controlled settings would help regulators and developers check how AI systems might suffice to generate or add harmful misinformation in situations when they are manipulating documents bought or generating sustainable reports. The sandbox testing would assist in the identification of applications that carry high-risk factors and offer further background practices and spur innovation in the technologies of any given detection and prevent platforms.

**Verified and safe communication channels to critical infrastructure:** In the case of supply chains and logistics, of particular focus is the use of safe and verified digital communication protocols that need to be promoted by policymakers and industry organizations. This involves the encouragement of the bills of lading, certificate of origin, and port authority announcements systems based on blockchains to ensure an immutable and audit trail record, which would make it much harder to inject AI-forged documentation.

These recommendations should be implemented through international cooperation in order to reduce the willingness among those participants for AI misinformation to go beyond the borders of different countries and allow the changes to be introduced at the regional level. The most promising one is a stratified system of rules, where technical criteria are integrated into education and control to avoid economic risks without curtailing the positive AI applications. Any other policy should be based on quantifiable results, and consistent measurement of such initiatives will assist in the continual enhancement of such initiatives as AI possibilities and deception approaches evolve in the future.

## 10.5 CONCLUSION

The economic effects of AI-generated misinformation in this chapter have been assessed in an organized way addressing five areas of concern; ESG reporting, international trade, global supply chains and port logistics, educational policy, and sustainable innovation in development. According to our findings, AI misinformation is not a hypothetical problem; rather, it is an economicallymeasurable interference with tangible effects on the market efficiency, policy effectiveness, continuation of operations, and development of technologies. The fact that regular statistical correlations were found in all sectors, which in-depth case study supports, recalls the systemic nature of the emerging issue.

In this work, AI-driven misinformation causes significant breaches in the reliability of ESG reporting, with each unit increase in the exposure being associated with the 42-percentage-point reduction in the reliability. As in the case of international trade, we observe that misinformation does not simply accompany the conflicts, but, in fact, is the cause of conflict, that each ten-point increase in levels of misinformation leads to a 3.8% rise in conflicts. Global supply chains and port logistics were most sensitive, with the highest disruption metrics per unit misinformation exposure increasing by 45%, making them a vital point of vulnerability. The susceptibility of educational systems to algorithm misinformation is quantifiable, and there is policy drift in regard to the empirical foundation of the postured evidence with exposure. And most importantly on the way to such requisite sustainability goals, our assessment shows disastrous lags in the adoption of green technologies due to misinformation.

The semiconductor trade rumor case study is an actionable instance of how the dynamics clarifies in the real world since, due to the production of the sophisticated content using AI, the market value of the stock dropped by 2.1 billion, and the supply chain was destabilized prior to the corrective action. The case and our additional findings indicate a list of structural vices in the contemporary economic systems – that is, the difference between the speed at which algorithmic false information is disseminated and the speed of the process in question being verified.

This chapter suggests some of the most important questions that should be posed in the future. To begin with, it will be required to conduct longitudinal studies to evaluate the possible fluctuation in the effect of misinformation due to further evolution of AI capabilities. Second, the comparative analysis of the alternative regulatory policies would be at a position to make recommendations on how the most suitable combination of policies would be found in various economic objectives. Lastly, an intrusion in industry-specific pre-emptive interventions and notifications, in particular, to supply chains and other logistical centers, can also possibly facilitate the anticipation and avoidance of the crises of misinformation, before a period when it can cause economic harm.

## REFERENCES


Adnani, E., & Haounani, A. (2024). Artificial intelligence in Morocco: Between ethics and regulation. *International Journal of Scientific Research (IRS-Journal)*, 2(3), 1234–1252.

Ammar, A., Amahmoud, H., El Menzhi, K., Mkik, M., & Hebaz, A. (2024, May). Critical Analysis of Logistics Transformation in Morocco under the Auspices of Artificial Intelligence: Government Perspectives From the perspective of Innovative and Technology Driven Public Management. In *2024, IEEE15th International Colloquium on Logistics and Supply Chain Management (LOGISTIQUA)*, Sousse, Tunisia (pp. 1–9). IEEE.

Atkinson, A., & Messy, F. A. (2012). Measuring financial literacy: Results of the OECD/International Network on Financial Education (INFE) Pilot Study. OECD Working Papers on Finance, Insurance and Private Pensions, No. 15.

Cole, S., Sampson, T., & Zia, B. (2011). Prices or knowledge? What drives demand for financial services in emerging markets? *Journal of Finance*, 66(6), 1933–1967.

Eriola, J. A., & Agbovoedo, E. C. (2023). Financial education and difficulties in accessing finance for women entrepreneurs in Benin. *African Journal of Management (RAG)*, 6(1), 17–30.

Hastings, J. S., Madrian, B. C., & Skimmyhorn, W. L. (2013). Financial literacy, financial education, and economic outcomes. *Annual Review of Economics*, 5(1), 347–373.

Hung, A. A., Parker, A. M., & Yoong, J. K. (2009). Defining and Measuring Financial Literacy. In: National Research Council*Measuring Financial Literacy: A Workshop Summary* (pp. 15–33). National Academies Press.

Jirari, S., Khiati, M., Mabchour, A., Ouboujema, E., Mohammed, B., & ElOmari, O. (2025). Artificial Intelligence and Morocco's Education System: A Strategic Tool for Enhancing Public Management and Institutional Performance. In *2025 International Conference on Circuit, Systems and Communication (ICCSC)*, Fez, Morocco (pp. 1–6). https://doi.org/10.1109/ICCSC66714.2025.11135414

Khiati, M., Elalaoui, A., Laator, A., Majd, A., Mkik, M., Tanane, O., & Imadi, I. E. L. (2025a). Modeling Sustainable Production Systems and Ecological Transition: The Key Role of Education Based on AI in the Green Economy. In: GonzálezVallejo, R., Moukhliss, G., Schaeffer, E., Paliktzoglou, V. (eds) *The Second International Symposium on Generative AI and Education (ISGAIE'2025). ISGAIE 2025. Lecture Notes on Data Engineering and Communications Technologies* (vol. 262, pp. 443–453). Springer. https://doi.org/10.1007/978-3-031-98476-1_35

Khiati, M., Jirari, S., Aissa, S., ELomari, O., Ouboujemaa, E., & Ali, B. (2025b). Artificial Intelligence and Smart Education as Dual Catalysts for Reinventing Global Trade, Entrepreneurial Innovation, and Intelligent Customs Logistics: A Multi-Layered Modelling Framework. In *2025 International Conference on Circuit, Systems and Communication (ICCSC)*, Fez, Morocco (pp. 1–7). https://doi.org/10.1109/ICCSC66714.2025.11135043

Khiati, M., Jirari, S., Metwalli, O., Rafi, N., Karim, A., Mkik, M., & Tanane, O. (2025c). Modeling an Intelligent Educational Ecosystem for Training Logistics Entrepreneurs via AI and Data Science. In: GonzálezVallejo, R., Moukhliss,

G., Schaeffer, E., Paliktzoglou, V. (eds) *The Second International Symposium on Generative AI and Education (ISGAIE'2025). ISGAIE 2025. Lecture Notes on Data Engineering and Com- munications Technologies* (vol. 262, pp. 343–352). Springer. https://doi.org/10.1007/978-3-031-98476-1_27

Landozi, S., & Hounkou, C. E. (2023). Influence of financial education on business performance in small businesses in Togo. *International Journal of Management Sciences*, 6(2), 37–39.

Lusardi, A., & Mitchell, O. S. (2011). Financial literacy around the world: An overview. *Journal of Pension Economics and Finance*, 10(4), 497–508.

Mandell, L., & Klein, L. S. (2009). The impact of financial literacy education on subsequent financial behavior. *Journal of Financial Counseling and Planning*, 20(1), 15–24.

Marouane, M., Salwa, M., & Hebaz, A. (2023, May). The Acceptance of Artificial Intelligence in the Commercial Use of Crypto-Currency and Blockchain Systems. In: Ranganathan, G., Papakostas, G. A., Rocha, Á. (eds) *International Conference on Information, Communication and Computing Technology* (pp. 163–177). Springer Nature Singapore.

Mkik, M., & Mkik, S. (2023, January). Acceptability Aspects of Artificial Intelligence in Morocco: Managerial and Theoretical Contributions. In: Motahhir, S., Bossoufi, B. (eds) *International Conference on Digital Technologies and Applications*, Fez, Morocco (pp. 65–74). Springer Nature Switzerland.

Mouatadid, N., Oukheda, M., Khiati, M., Saile, R., & Kettani, A. (2025). Determinants of aged cheese consumer preferences in Morocco—A cross-sectional study of economic, cultural, and social factors influencing purchasing behaviors. *Frontiers in Nutrition*, 12, 1600873. https://doi.org/10.3389/fnut.2025.1600873

Mouhsine, S., Haouat, A., Amiri, W., & Elabid, S. (2023). Smart logistics in Morocco: What future?. *International Journal of Accounting, Finance, Auditing, Management and Economics*, 4(2-2), 136–163.

Mouzoun, Z., & Ammi, A. (2023). The Repercussions of Financial Exclusion in Morocco on the Environment: Analysis through the Application of Rooted Theory. In *SHS Web of Conferences* (Vol. 175, p. 01035). E3S Web of Conferences, EDP Sciences.

Ndiaye, M. (2024). For an endogenous and inclusive approach to Artificial Intelligence in Africa.

Roussi, K. (2022). Artificial Intelligence at the service of the chartered accountant's profession. *International Journal of the Researcher*, 3(2), 14–17.

Safia, B., & Ahlam, C. (2024). Artificial intelligence AI is transforming the business world. *Journal of Contemporary Business and Economic Studies*, 7(1), 15–25.

Ullah, S., Nobanee, H., Azim, F., & Khiati, M. (2025). Financial foundations for sustain-ability: How business sophistication, tax policy, and technology shape ESG in belt and road initiative countries. *Global Sustainability*, 8, e3. https://doi.org/10.1017/sus.2025.4

Part IV

# Ethical challenges and future trends

Chapter 11

# Role of data protection law in combating generative AI misinformation

## A visible threat to privacy

*Showkat Ahmad Wani and Renuka Jaggi*

## 11.1 INTRODUCTION

In the contemporary world, advanced technologies like artificial intelligence (AI) can generate different content in a textual, image, or video form. Nevertheless, such a technology is the need of an hour, but AI invites issues and concerns of privacy breach and misinformation too. It is imperative to analyze the efficacy of existing data protection laws as a measure to protect security threats, particularly AI-sourced misinformation risks. To understand the data-related rights, the study further examines theoretical concepts in law, including personal autonomy, the right to be forgotten, and public law, as mentioned in John Stuart Mill's harm principle and John Rawlsian justice theory (Huq, 2013). Comparative case studies are offered in order to emphasize the ways in which generative AI tools have been used for in political campaigns, public health disinformation, and impersonation-based fraud. The research also performs an analysis to compare the legal answers of three jurisdictions, i.e., the European Union (EU), India, and the United States. While both the EU's General Data Protection Regulation (GDPR) and AI Act provide a framework and risk-based method, India's DPDPA lacks strong enforcement capabilities. The United States still does not have uniform legislation to control AI-facilitated disinformation. Throughout all jurisdictions, the study identifies such gaps as a lack of obligations for AI developers, weak cross-border enforcement, and poor synthetic content recognition under present data laws. In a digital age, dominated by transformation, the introduction of generative AI is an engineering phenomenon and an administrative difficulty. From realistic deepfakes to news articles written by AI, generative AI has revolutionized the way information is produced and consumed. But with such power comes an enormous risk, the erosion of privacy and public interest through disinformation. The widespread dissemination of AI-generated false narratives, which are frequently indistinguishable from the actual content, represents a major threat to democratic institutions, personal autonomy, and social stability. The major challenge is data privacy through AI systems because generative AI is based on large datasets where such data contain personal or sensitive information, which

DOI: 10.1201/9781003626558-15

poses significant privacy risks. The Cambridge Analytica case and the use of deepfake technology in political campaigns are the best examples of how AI-based disinformation can destabilize the public sphere and intrude into private data rights (Cadwalladr & Graham-Harrison, 2018).

According to Daniel Solove, "Privacy is perhaps the most critical concept of our era, but also one of the most elusive. While advancing technology continually makes information more accessible, academics, activists, and lawmakers have grappled with how to define privacy, most admitting that the task is all but impossible" (Solove, 2008). These technologies are often trained on publicly available data from the internet, which creates a concern over informed consent, data ownership, and intellectual property (Westin, 1967). The conflict between technological innovation and legal regulation is, therefore, stark. The policy and legal discussion of generative AI have enhanced in the last few years (Solove, 2002). The Council of Europe, Organisation for Economic Co-operation and Development (OECD), and United Nations have all published policy documents highlighting the necessity of ethical AI as well as robust data governance structures (OECD, 2019). In reply, a number of nations have begun to develop or revise their data protection legislation. There are still problems with the existing laws because it was developed to ensure protection in case of AI-generated data breaches.

The traditional legal mechanism is still not framed to provide solutions to the new challenges of deepfakes and AI-generated misinformation. Beyond this, there is an urgent need to differentiate between misinformation and disinformation. Misinformation is false or misleading information given without any intention to deceive, whereas disinformation is intentionally false information given to be misleading (Wardle & Derakhshan, 2017). They can both be employed with generative AI. A generative model either accidentally generates false content as a result of flawed training data, or it can be exploited by bad actors to create and disseminate disinformation operations at scale. It becomes imperative to explore the convergence of data protection law and generative AI and how it can be used to fight misinformation while protecting privacy and the public interest.

The question arises whether conventional data protection legislation adequately addresses the challenges of generative AI–driven misinformation through powerful AI-focused regulations. By analyzing important legal tools like the European Union's GDPR, India's Digital Personal Data Protection Act (DPDPA) 2023, and other global measures, this article assesses the effectiveness of legal safeguards against the threat of disinformation and data exploitation. Section 11.2 offers a conceptual framework of generative AI and its spread of disinformation (European Parliament & Council of the European Union, 2016). Section 11.3 outlines and contrasts data protection regimes in different jurisdictions (European Parliament & Council of the European Union, 2016). Section 11.4 identifies privacy and public interest threats. Section 11.5 points out the gaps in the law and challenges to regulation. Section 11.6 sets out suggested legal and policy measures (European Parliament & Council of the European Union, 2016).

## 11.2 CONCEPTUAL FOUNDATIONS: GENERATIVE AI, MISINFORMATION, AND PRIVACY

Generative AI is a machine learning algorithm that recognizes patterns or classifying data, can create completely new instances of data. They encompass models such as generative adversarial networks (GANs), variational autoencoders (VAEs), and transformer-based models such as GPT and BERT. They can synthesize photorealistic images, coherent text, simulated speech, and even synthetic biometric data. The fundamental concept underpinning generative AI comes through the training of models with enormous datasets so that they can acquire language, visual, or behavioral patterns. These machines function on probabilistic learning, anticipating the next most probable element in a sequence or filling in missing information given prior observed patterns (Goodfellow et al., 2014). This makes them phenomenally powerful but troublesome, especially without constraints on derivation and the nature of training data.

Both misinformation and disinformation are interrelated. While misinformation means anything that is inaccurate or misleading irrespective of whether done with intent or without intent, disinformation, on the other hand, means deliberate dissemination to deceive general masses. AI is complicated because it can easily create false information in various forms like images and videos without malicious intent (Wardle, 2017). However, we should understand that AI-generated content without malicious intent is dangerous if not verified. It can easily create confusion, which ultimately erodes public trust.

AI can easily generate private data of the "Users." It has the potential to extract personal information through social media and past activities of the "User" profiles (Zuboff, 2019). It is a hyper-personalization that creates imminent threats that are referred to as "cognitive hacking," where false information is spreading because of algorithmic influence (Kreps & Kriner, 2023). This AI-generated disinformation not only manipulates public thinking, but it also directly impacts political and economy manipulation, which is not just psychological but also political, economic, and public health at large. The generative AI access data are publicly available; moreover, anonymized data can also be extracted with the help of other relative information of other data. Such AI-generated content raises individual privacy issues because it violates a person's expectation of anonymity and consent (Narayanan & Shmatikov, 2008).

Technological inventions have made life easy. In contemporary times, most of the work is done online, and the dependence on online platforms has increased significantly. Diverse online platforms are used for various purposes, including learning, business, entertainment, and socialization. The evolving information age is transforming our lives in ways we could never imagine. In a data-based world, privacy is the first thing that faces frequent security threats. The processing of data, voluntary data sharing by people in social media, commercialization of personal data, etc., are serious challenges for the privacy rights of the individuals. As we know, privacy is an inevitable and inalienable right, and the recognition of privacy as a fundamental human right needs to be protected from generative AI systems (Westin, 1967).

Today, it is very difficult to recognize which content is real or AI-generated fake. AI content mostly deceives the public by providing fake information. Deepfakes are very dangerous because AI can manipulate the audios and videos easily and can portray it in a real way (Chesney & Citron, 2019). The victims of deepfake find it difficult to seek legal remedies, particularly jurisdiction issues. There is still a legislative gap to address how to ensure safety of personal rights and accountability of false information.

Another significant challenge is the transparency of AI-generated content because these models operate as "black boxes." The "black box" functions in an AI system, and its internal workings is a mystery to its users. "Users" can see the system's inputs and outputs, but they can't see what happens within the AI tool to produce those outputs. This makes harm attribution difficult. If a biased dataset has been utilized in training an AI system that generates discriminatory content or misinformation, liability is a time-consuming effort to determine. Who would be held responsible, the developer, the data provider, or the platform that made the output accessible? Worldwide, answers to these challenges are emerging (Mittelstadt et al., 2016). The EU's AI Act is groundbreaking in proposing the definition of "high-risk AI" as being targeted at those systems with the potential to impact in-depth substantive rights. Even that does not address the deeper cognitive challenge of how to distinguish AI-generated truth from falsehood and how to integrate ethical judgment within probabilistic models (European Commission, 2021).

Although, people also support an "AI ethics by design" strategy in which ethical concerns are incorporated into the model's planning rather than added later by means of regulation. Others demand even stringent data protection standards with the right to explanation, correction, and redress principles already enshrined in the GDPR but haphazardly applied across jurisdictions (Floridi & Cowls, 2019). Public interest is paramount for stability in a society. AI-generated contents create cognitive confusion, and eventually in this process, truth and evidence lose their authority. Therefore, this part has outlined three interconnected ideas, i.e., generative AI as a machine for creating content, misinformation as danger fuelled by that machine, and privacy as the most vulnerable ground. The convergence of these issues represents a legal and ethical imperative for anticipatory governance, open liability structures, and stronger privacy safeguards to protect the people and society in general.

## 11.3 LEGAL DOCTRINES AND THEORETICAL FRAMEWORKS

The law of managing generative AI disinformation comprises multiple strands of legal and philosophical thought, primarily concerning privacy, autonomy, and public interest. Here, the central legal rules and theoretical contexts supporting contemporary strategies on AI-generated disinformation and privacy violations are necessary to resolve.

### 11.3.1 Data protection as a fundamental right: comparative overview

Within most democratic jurisdictions, the data protection right has evolved from statutory affirmation into a constitutionally embedded right. The EU has led this development, and the Charter of Fundamental Rights of the European Union named data protection as a separate fundamental right of the right to privacy (European Union, 2000, art. 8). Article 8 of the Charter categorically guarantees protection of personal data, consolidating the wider application of the GDPR.

India replicated through judicial intervention in the case of Justice K.S. Puttaswamy case, and the Indian Supreme Court recognized the right to privacy as a fundamental right under Article 21 of the Constitution (Supreme Court of India, 2017). While the Digital Personal Data Protection Act, 2023 does not explicitly state data protection as a fundamental right, the interpretation of the Supreme Court requires a high threshold for data collection and processing.

In comparison, the United States has a distant model of data protection. No single federal law ensures an overall right to data protection. Rather, legislations such as the Health Insurance Portability and Accountability Act (HIPAA) and the California Consumer Privacy Act (CCPA) deal with data privacy in a disorganized fashion (Schwartz & Solove, 2011). The U.S. Supreme Court has long placed privacy within the Fourth Amendment paradigm, focusing on protection against state surveillance but with no overarching framework for corporate abuse of data.

### 11.3.2 Informational self-determination and the right to be forgotten

Informational self-determination is the ability of individuals to decide about the disclosure and utilization of their personal data. It came into focus post the 1983 German Federal Constitutional Court's census judgment, where it was emphasized that citizens should be able to control what information about them is being gathered and for what reason (Bundesverfassungsgericht [Federal Constitutional Court], 1983).

This principle forms the foundation of most rights of the GDPR, such as data access, correction, and erasure. One important expression is the "right to be forgotten," enshrined in Article 17 of the GDPR (European Union, 2016b, art. 17). This right entitles a person to ask for personal data to be deleted when no longer needed or upon withdrawal of consent. While contentious, particularly in reconciling public interest and freedom of expression, this doctrine is instrumental in improving the effect of AI-generated content that defames or misrepresents persons. India's legal system at present lacks an effective right of being forgotten, even though some High Courts have ruled in cases for delisting material from search engines (Sahoo, 2021). There is still conflict between data protection and free speech in Common Law Nations including the USA, irrespective of the First Amendment in their Constitution.

### 11.3.3 Doctrine of data minimization, purpose limitation, and lawful processing

Three doctrines, all connected with each other, form the core of contemporary data protection law, i.e., data minimization, purpose limitation, and lawful processing. These ideas are codified in Article 5 of the GDPR and reiterated in India's DPDPA and other international instruments such as Brazil's LGPD.

- Data minimization simply requires that only data required for a stated purpose must be gathered.
- Restriction of purpose means that data must be processed only for the originally specified purpose when collected.
- Legal processing means data processing must comply with explicitly defined legal bases, including consent or public interest (Tufekci, 2014).

These are all at odds with the current operation of generative AI systems. Model training is generally done with massive sets and minimal concern for a particular purpose or user consent. Anonymized data can easily be fed to generate outputs that breach privacy or disseminate misinformation, undermining the legal foundation of such processing as it currently stands under the existing law.

### 11.3.4 Public interest law theories: Mill, Rawls, and Habermas

Public concern to regulate technology has its roots in classical as well as contemporary political theory. John Stuart Mill's "harm principle" assumes liberty can be curtailed only to avoid harm to other individuals. Falsehood that leads to harm to reputations or the undermining of elections obviously amounts to such harm, warranting legal intervention even under liberal democracies (Mill, 1859). John Rawls theory of justice, i.e., the "difference principle," enforces it as an obligation that institutional and social arrangements are advantageous to the less wealthy. In the case of AI, it supports regulation on behalf of groups who are disproportionately affected by misinformation and data abuse (Rawls, 1971). Jürgen Habermas public sphere theory anticipated democratic legitimacy through rational communication among citizens (Habermas, 1989). Generative AI disinformation destroys this foundation based on communication, creating a debate on environment and distortion. Public interest law, as a result, has a duty to preserve the communicative conditions for democratic engagement, including reliable and correct information.

### 11.3.5 Algorithmic accountability and legal personhood

The development of autonomous systems raises new questions of liability and accountability. Algorithmic accountability refers to the obligation on the part of the developers and deployers of AI to deliver transparency, fairness, and contestability in decisions made by machines (Pasquale, 2015). The current legal framework assigns liability to natural or legal persons, not to machines. With increasingly autonomous AI systems, however, the debate over "electronic personhood" has increased momentum in European policy discourse (European Parliament, 2017).

The critics argue that granting legal personhood to AI would accelerate human irresponsibility and provide corporate defenses to wrongdoing (Pagallo, 2013). However, the imperative for transparency in attributing responsibility in cases of misinformation persists. Courts and legal regulations need to establish elaborate standards for tracking causation and accountability in multi-agent AI systems.

### 11.3.6 Legal v. ethical regulation of AI-generated content

The ethics presents flexible principles like fairness and autonomy, whereas law has enforceable rules. Most of the organizations have embraced ethical standards, but these are typically on a voluntary basis and have no legal binding. Ethical frameworks also differ greatly from culture to culture, resulting in irregular implementation (Jobin, Ienca, & Vayena, 2019).

Legal governance, in comparison, guarantees a level playing field of rights and duties. It also tends to fall behind technology development. Closing this gap is achievable via a hybrid model, described by some as "soft law," whereby ethical norms are increasingly codified in enforceable standards. Herein lies the challenge of ensuring such standards are non-symbolic yet effective in improving actual harms caused by generative AI (Fenwick, Siems, & Vermeulen, 2009).

## 11.4 CASE STUDIES OF MISINFORMATION VIOLATIONS OF GENERATIVE AI

Investigating real cases of generative AI misuse reveals the actual shortfalls in current data protection mechanisms in stopping misinformation and maintaining privacy. The following case studies identify the diverse industries where generative AI has caused adverse social and legal challenges, particularly without proper legal systems for evading accountability.

### 11.4.1 Deepfakes and electoral misinformation: the 2024 elections

Prior to the 2024 general elections in India and America, deepfake technology was used to circulate artificial videos mimicking political candidates. In India, a deepfake of a local politician was circulated on WhatsApp, making it seem like he uttered abusive statements on communal issues. Despite quick denials, the video went viral, resulting in reputation damage and increased tensions within the electoral process (Barari, Lucas, & Munger, 2025). In the United States, even, a deepfake voice recording with President Biden's impersonation called on New Hampshire voters to stay away from the primary election, which led Federal Communications Commission investigations to initiate investigation (Associated Press, 2024).

These instances expose a regulatory vacuum. Although there are laws against misinformation in India's IT Act and the U.S. Federal Election Commission oversees campaign advertisements, both systems do not adequately prepare for the threats raised by deepfakes. Further, current data protection legislation fails to tackle the impersonation of people or the artificial copying of their biometric information, like voice or facial attributes.

### 11.4.2 AI-generated health misinformation

AI-generated content greatly aided health misinformation during the COVID-19 pandemic. Chatbots that were trained on unverified information created untrustworthy advice regarding preventive activities, vaccine effectiveness, and treatment guidelines (Xie, 2025). A chatbot based on GPT was once used to answer a question about vaccines with false information, indicating that mRNA vaccines change DNA. The misinformation was subsequently referenced and compounded on conspiracy-fueled platforms.

Such violations not only risk public health but also demonstrate the shortfall of data protection laws. The GDPR and India's DPDPA primarily address consent and processing but don't include liability on platforms creating or spreading hateful content with the aid of AI. Further, existing legislations lack adequate means to check or recall AI-generated content, resulting in mass disinformation.

### 11.4.3 Synthetic media and privacy violation in celebrity or revenge porn situations

Without consent sharing of sexually explicit content, deepfakes is among the worst privacy violations enabled by generative AI. Many instances have been reported, where public figures, especially female celebrities, have had their faces superimposed over pornographic material. In one high-profile case in 2023, unauthorized deepfake photos of a Bollywood star went viral on social media, with demands for the platforms hosting such content to face intensified legal consequences (Pavis, 2021).

Though there are some jurisdictions, such as the United Kingdom, which have enacted laws criminalizing deepfakes of pornography, most data protection frameworks don't classify synthetic content as per se personal data, thus leaving loopholes. Such content proves difficult for victims to have deleted, particularly when it is hosted by servers in foreign jurisdictions.

### 11.4.4 Chatbots used for fraud and impersonation

Chatbots powered by AI have also been misused for deceptive and fraudulent activities. In 2023, some banks in Southeast Asia and Europe had reported phishing attacks where chatbots were employed to pose as customer care representatives, seeking sensitive financial details from users (Brewster, 2021). In India, there were reports of scam calls where AI voices were used to impersonate relatives to get money from innocent victims.

These events illustrate the shortcomings of existing cybersecurity and data protection regulations. Even if financial legislations provide for mechanisms of compensation, liability attribution remains unclear when dealing with AI systems. In addition, the existing data protection legislation does not typically mandate proactive auditing of chatbot responses or the training data sets behind such systems.

### 11.4.5 Analysis: the failure of current legal frameworks

Throughout these case studies, some common themes reappear. First, current data protection legislations are back-end focused and not pre-emptive. They do not provide hints to foresee how generative AI can be abused. Second, the legal environments tend to be narrowly centered on controllers and processors of data and do not cater to personal roles like model builders, trainers of AI, and distributors of content. Third, most laws do not look at synthetic data content that is created without literally imitating individual information, as coming under the scope of data protection.

### 11.4.6 Societal, legal, and regulatory responses

To these issues, governments, international organizations, and civil society have begun to react. The EU's draft AI Act establishes obligations for high-risk AI systems, including transparency requirements and human review (European Commission, 2021). India's DPDPA 2023 contains provisions for penalties in the event of noncompliance but remains quiet regarding harms specific to AI. In the United States, several state legislatures, including California and Illinois, have begun regulation of synthetic impersonation and biometric data through targeted legislation.

Social media platforms have also introduced labeling of AI-generated content, and the regulatory mechanism is typically implemented irregularly. Civil society efforts at public education and fact-checking try to counter the misinformation provided by AI. But since there is no mechanism for enforcing universal standards, the efforts are disconnected and unsuccessful.

Therefore, in raising the awareness about the danger of generative AI disinformation, legal and institutional measures must be taken but are insufficient. An urgent need is a firm international framework for data protection interests and calibration of AI technologies.

## 11.5 COMPARATIVE LEGAL ANALYSIS: EUROPEAN UNION–GDPR AND THE AI ACT

The EU has also been a front-runner in setting data protection norms in the world with the GDPR, and more recently the Artificial Intelligence Act (AI Act) of 2024. These instruments form the building blocks of the EU's regulatory scheme for dealing with privacy risks and public interest of generative AI.

### 11.5.1 Overview of GDPR provisions relevant to AI

Articles 5–9 of the GDPR establish an organized framework for legal data processing. The fundamental principles are prescribed by Article 5, which includes lawfulness, fairness, transparency, limitation of purpose, data minimization, accuracy, limitation of storage, integrity, and confidentiality (European Union, 2016c). They are generic principles but do directly apply to how AI systems must be designed and used, specifically if personal data are involved.

Article 6 defines lawful bases for processing personal data, such as consent, performance of a contract, legal obligation, protection of vital interests, public interest, and legitimate interests. Article 9 proposes more stringent protections for special categories of data such as health, race, and biometric data, which are typically used or inferred by generative AI systems (European Union, 2016c).

Article 22 is especially applicable. It entitles people to have the right not to be made subject to a decision based solely on automated processing, such as profiling, which has legal or similarly affecting effects (European Union, 2016c). This article directly overlaps with the results of generative AI, particularly where such systems operate in employment, credit, or social service decisions.

### 11.5.2 The right not to be subject to automated decision-making

The GDPR's Article 22 offers a rights-based paradigm against AI-reliant decisions that are unmitigated by human input. This is especially necessary where AI-generated misinformation may affect legal rights like losing a loan based on an AI classification error. But its application is limited; exceptions apply if a decision is needed for a contract, provided by law, or on express consent. These exceptions introduce uncertainty in applying the rule to generative AI output that, although not "decisions" in the strict sense, may have similar real-world consequences (European Union, 2016c).

### 11.5.3 AI act and classification of risk categories

The EU Artificial Intelligence Act divides AI systems into four risk categories: unacceptable, high-risk, limited-risk, and minimal-risk. Generative AI systems are deemed to be general-purpose AI for the most part, though they can fall under the high-risk category if used in sensitive fields such as health, law enforcement, or education (European Union, 2024).

High-risk systems need to be subjected to strong obligations like risk management, data governance, transparency, human oversight, and accountability measures. Documentation and conformity assessment are mandated under the Act, and this creates a pre-market testing administration in sharp contrast to the afterward enforcement mechanism in other jurisdictions (European Union, 2024).

Most significantly, the AI Act places obligations on "foundation model" providers, like the developers of large language models (LLMs) such as GPT, to report training data, testing procedures, and safeguards against undesirable outputs, including disinformation. It makes an important shift in the necessary regulations of generative AI in contemporary times (European Union, 2024, Art. 52–54).

### 11.5.4 Data protection by design and data protection impact assessments (DPIAs)

According to Article 25 of the GDPR, data controllers must introduce "data protection by design and by default." It implies that data protection principles must be built into the design and planning of processing systems, such as AI algorithms (European Union, 2016d).

Besides, Article 35 requires Data Protection Impact Assessments (DPIAs) where processing is expected to be of high risk to the rights and freedoms of individuals. This applies primarily to generative AI systems that can produce personalized misinformation or impersonating a person. DPIA's for

AI systems recommended by the European Data Protection Board (EDPB) include assessments of data quality, bias, and explainability (European Data Protection Board, 2019).

### 11.5.5 The role of the European Data Protection Board (EDPB)

The EDPB is at the forefront of setting the tone for how the GDPR applies to new technologies, such as AI. It has produced many of these guidance documents and opinions detailing how GDPR values apply to algorithmic decision-making, profiling, and the processing of biometric data (European Data Protection Board, 2018).

In 2021, the EDPB and European Data Protection Supervisor (EDPS) demanded prohibition on AI systems employing biometric identification in public areas based on growing fears of loss of privacy. The EDPB also proposes cross-border enforcement and harmonizes the application of the GDPR across Member States, providing a key institutional check on regulation of AI-based disinformation (European Data Protection Board & European Data Protection Supervisor, 2021).

### 11.5.6 Gaps and effectiveness in combating AI misinformation

Even with its comprehensiveness, however, the GDPR and AI Act retain several major weaknesses in tackling the dangers of generative AI misinformation (Table 11.1).

*Table 11.1* European data protection law and generative AI regulation

1. Non-personal data loophole: Artificially created content by AI that fails to literally reproduce personal data mostly evades GDPR control, although it impersonates or deceives.
2. Narrow scope of Article 22: The safeguards for automated decision-making do not explicitly apply to AI-created misinformation unless it has a direct link to a legal ruling.
3. Insufficient enforcement capabilities: Data protection authorities in most Member States lack the financial means and technical expertise to verify advanced AI models (Kuner et al., 2017).
4. Uncertainty over content moderation: The GDPR does not explain intermediary responsibility or platform responsibility for hosting generative AI-generated material which constitutes disinformation.
5. Delayed enforcement: The AI Act includes transition periods and will not be fully enforceable until 2026, leaving a regulatory gap in the interim.

### 11.5.7 Digital personal data protection act 2023 (DPDPA) India: comparative analysis

Until 2023, India did not have a separate law on data privacy. The data protection issues were governed by the Information Technology Act, 2000 (IT Act) and notified rules. There was a breakthrough in 2017, when the Supreme Court of India in the Justice K. S. Puttaswamy (Retd.) vs Union of India case upheld that privacy is a fundamental right. This led to the formulation of a comprehensive data protection framework for India. The (DPDPA) 2023 is a milestone step that protects privacy of individuals. Its main objectives are to protect data-related rights, particularly related on consent, and data minimization.

### 11.5.8 Key features and consent architecture

Primarily, the data protection law in India embeds the concept of "consent-based" processing of data. Section 4 mandates that personal data should be processed only for legal purposes with the data principal's consent or for lawful purposes, as defined under the Act (Government of India, 2023, § 4). Free, specific, informed, and unambiguous consent is needed, mirroring international best practices under GDPR.

The Act provides for the notion of "Data Fiduciaries," which are bound to handle data transparently and responsibly (Government of India, 2023, § 7). Yet, categorization of fiduciaries into "Significant Data Fiduciaries" considering the volume and sensitivity of handled data has also caused concerns regarding potential regulatory inequalities. Further, the consent architecture assumed by the law does not provide for cases with AI-produced data in which no direct collection from the person took place. This causes a basic mismatch when used in generative AI systems that learn from large-scale public databases without having explicit individual permission (Government of India, 2023, § 10).

### 11.5.9 Lack of mechanisms for meaningful enforcement

Although the DPDPA provides the Data Protection Board of India (DPBI) as the main enforcement body, it is structured to lack structural independence. Section 19 provides for the appointment of Board members by the Central Government, which would defeat the aim of institutional autonomy (Government of India, 2023, § 19). Additionally, the Act confers the government with authority to exempt some government agencies from its operation under Section 17, opening ample space for state intrusiveness and diluting public trust (Government of India, 2023, § 17).

The lack of binding deadlines for grievance redressal and absence of class-action provisions also limits the force of enforcement. Further, the Act does not provide provisions for mandatory algorithmic audits, explainability standards, or transparency requirements, all key weapons in fighting back against generative AI disinformation.

### 11.5.10 Insufficiencies in regulating cross-border data flows and AI

DPDPA's cross-border transfer solution is significantly diluted compared to that of the GDPR. Under Section 16, notification by the Central Government of states to where personal data may be exported is provided for (Government of India, 2023, § 16). It does not have objective criteria of adequacy and does not require reciprocal protection.

This is particularly difficult in the case of generative AI models developed or hosted elsewhere but trained on Indian data. No local storage, data localization, or contractual terms for cross-border processing are required. The Act also lacks express provisions to handle obligations of creators of AI, operators of platforms, or intermediaries in handling misinformation generated by AI tools.

### 11.5.11 Function of the Data Protection Board of India (DPBI)

The DPBI is empowered to investigate grievances, levy fines, and order directions for compliance. Its dependence on the executive, as mentioned, undermines its power (Government of India, 2023, § 19 {2}). Another important question is regarding powers of investigation which it does not have *suo motu* investigation like the European Data Protection Board and technical advisory departments for AI-related matters.

The Board's adjudicatory powers are also limited to contraventions of the DPDPA. It is unclear whether misinformation harms caused by AI outputs that do not involve direct personal data breaches fall within its jurisdiction. As generative AI increasingly produces content that misrepresent the boundary between synthetic and personal data, the DPBI's lack of technological mandate poses a significant enforcement dilemma.

## 11.6 CHALLENGES IN ALIGNING WITH THE CONSTITUTIONAL RIGHT TO PRIVACY

The milestone judgment in Justice K.S. Puttaswamy (Retd.) v. Union of India (2017) held that privacy is a constitutional right enshrined under Article 21 of the Indian Constitution. It laid down the criteria of legality, necessity, and proportionality as the condition precedent for any invasion of privacy (Supreme Court of India, 2017).

The sweeping exemption of government bodies by DPDPA, its ambiguity regarding retention of data, and absence of judicial remedies circumvent proportionality and legality tests upheld by the Supreme Court. Moreover, its exclusion of rights such as the right to be forgotten, right to object to automatic processing, and right to explanation, recognized in GDPR, points toward a weaker protection mechanism (Supreme Court of India, 2017). The mental privacy, personal reputation, and autonomy face significant threats from generative AI. There is an urgent need to have a comprehensive approach to protect rights of the individuals. Although India enacted a data protection law, i.e., DPDPA, to govern digital platforms, but there are existing jurisdictional enforcement challenges and cross-border regulations. The existing law on data protection needs to be revised and amended align with constitutional privacy jurisprudence to incorporate AI-specific provisions.

## 11.6.1 United States: sectoral approach and emerging regulations

The USA has a separate regulation for different sectors as compared to broader and inclusive regulations enacted by the EU and India. While the system offers flexibility, it is not the uniformity needed to combat the cross-cutting generative AI and disinformation threats.

### *11.6.1.1 Sector-specific laws and patchwork protection*

There is no comprehensive federal data protection legislation in the United States. Instead, privacy law is scattered in sectoral acts like the Health Insurance Portability and Accountability Act (HIPAA) for health data, the Children's Online Privacy Protection Act (COPPA) for children, and the Gramm–Leach–Bliley Act (1999) (GLBA) for financial data. While these legislations are effective within their domains, they fail to confront the cross-sectoral and general-purpose character of generative AI tools (Schwartz, 2004).

The most advanced state-level privacy laws are the California Consumer Privacy Act (CCPA) and the subsequent California Privacy Rights Act (CPRA). They offer rights like access, correction, erasure, and opt-out of sale of data. These laws still lack provisions particularly addressing AI-generated content or disinformation (California Legislature, 2018).

## 11.6.2 Lack of federal AI-specific privacy regulation

Federal agencies such as the Federal Trade Commission (FTC) and the National Institute of Standards and Technology (NIST) have issued guidance on AI, but there is no federal AI legislation governing the training, deployment, or accountability of generative models. The FTC has employed

Section 5 of the FTC Act to act against deceptive or unfair AI practices, but it is a retrospective approach ill-suited to stopping AI misinformation at scale (Federal Trade Commission, 2023).

The NIST's Framework for AI Risk Management (2023) encourages responsible development and use of AI through an emphasis on fairness, transparency, and responsibility. It is, however, voluntary and lacks enforcement ability. Direct restrictions on data use or the spread of misinformation are not established by the framework; therefore, a massive regulatory chasm exists (National Institute of Standards and Technology, 2023).

### 11.6.3 Emerging legislative proposals and limitations

Several legislative bills have been introduced to fill this legal gap to protect data. The Algorithmic Accountability Act, reintroduced in 2023, seeks to mandate impact assessments of AI-driven decision-making systems. The American Data Privacy Protection Act (ADPPA) seeks to consolidate state laws and offer basic federal protections, but has been stalled in Congress (U.S. Congress, 2023).

Moreover, none of these proposed bills directly reference generative AI. They target transparency and consent but do not mandate AI-specific safeguards such as training data audits, misinformation prevention strategies, or liability for synthetic media-induced harms (Citron, 2018).

## 11.7 CHALLENGES TO REGULATING GENERATIVE AI MISINFORMATION

The sectoral model is highly challenging in addressing generative AI misinformation:

1. Regulatory fragmentation allows AI developers to exploit jurisdictional variations, escaping intensive examination.
2. Section 230 of the Communications Decency Act platform immunity shields intermediaries from liability for claims based on user-generated content, making it challenging to act against the dissemination of AI-enabled disinformation (United States Code, 1996).
3. Lack of special statutes addressing misinformation implies that most harms will have to be dealt with under tort or consumer protection statutes, which are not suitable for imminent harms.

### 11.7.1 Judicial and state-level interventions

American courts have not developed clear jurisprudence regarding liability for AI-created disinformation. Defamation and privacy torts need evidence of actual malice or overt harm criteria hard to meet with anonymized or artificial content (Prosser, 1960).

State legislatures are increasingly taking initiative. Illinois' Biometric Information Privacy Act (BIPA) has been used in cases involving facial recognition and deepfakes. Texas and Virginia have enacted laws requiring disclosures when synthetic media is used in political advertisements. While promising, these measures are disorganized and lack the scale to combat nationwide harms (Virginia Code § 24.2–956).

### 11.7.2 Regulatory agencies and oversight capacity

The FTC remains the primary enforcement agency for deceptive AI practices. In 2023, it opened investigations into AI tools falsely impersonating human users and issued warnings about AI-generated content affecting elections and healthcare. However, the agency lacks technical capacity, financial resources, and statutory mandate to regulate generative AI comprehensively (Federal Trade Commission, 2023).

The U.S. Department of Justice and Cybersecurity and Infrastructure Security Agency (CISA) also raised concerns about the national security consequences of AI misinformation, particularly in electoral interference and biothreats. Nonetheless, these institutions lack civil enforcement powers like EU DPA's (CISA, 2023).

The United States patchwork data protection environment and lack of a binding AI-specific regulation in place substantially impede its capacity for responding to the privacy and disinformation threats of generative AI. Nevertheless, the federal guidelines and state level measures are effective but are providing limited safeguards. To make it more robust and comprehensive, the federal legislation becomes necessary to enhance data protection laws from traditional laws of privacy which will deal with emerging deepfakes and misinformation.

### 11.7.3 Cross-border nature of generative AI platforms

We have to understand that generative AI programs are not a country-based problem. The AI models that create image generation programs and language models are trained in such a way to run across jurisdictions. Suppose a chatbot developed in the USA was trained using data collected from European users but used for marketing to the Indian market, regarding what law is to apply, enforcement jurisdiction, and rights of data principles (Veale & Edwards, 2018).

There is currently no global legal instrument addressing AI-specific cross-border data processing. While the GDPR includes provisions on international data transfers under Chapter V, (European Union, 2016a), other jurisdictions like India's DPDPA rely on executive discretion (Government of India, 2023). The lack of consistency and mutual recognition agreements across countries creates a vacuum that AI developers can exploit.

Without binding global standards or interoperability protocols, addressing misinformation on a platform operating beyond a single country's boundaries remains elusive (Ashutosh, 2024).

### 11.7.4 Legal lag: AI evolutionary pace vs. regulatory catch-up

Generative AI evolution is far ahead of the legislative processes of most nations. Laws take years to prepare, debate, and apply. New AI models come out in months or weeks. This discrepancy results in technologically new problems being addressed by regulatory tools outdated in conception (O'Sullivan, 2023).

Furthermore, courts have been behind the curve. There are *meager case laws* and a largely untested precedent regarding AI-generated disinformation, particularly regarding liability, standing, and evidence (Zuboff, 2019). There are also regulatory sandboxes and test-bed managements yet to mature, leaving room for proactive regulatory mechanisms. The technology cannot be stopped but can be accounted and regulated due to the privacy challenges (Guimarães Moraes, 2024).

### 11.7.5 Fragmentation and lack of harmonization internationally

Worldwide regulation of generative AI is characterized by fragmentation. The EU paces with the GDPR and AI Act, (European Union, 2024), India has enacted the DPDPA, and the USA follows a sectoral approach. At the same time, Global South countries are unable to develop coherent AI regulatory schemes, as is often the case from capacity or resource constraints (Bincoletto, 2021).

There are existing ambiguities in tackling global regulations on technology-driven AI models. The regulations are different in every country that led to confusions of jurisdictional issues because developers operate from different countries. The lack of international harmonization of the law also hinders cross-border enforcement of privacy rights and takedown notices for misinformation (Roberts et al., 2024). Initiatives such as the OECD AI Principles and UNESCO's AI ethics guidelines provide useful frameworks but are not enforceable (OECD, 2019).

### 11.7.6 Weak penalties and lack of state capacity

Even where such laws are in place, compliance is an enormous barrier. Data protection agencies across most countries are ill-funded, inadequately operated, and lacking in technical capabilities to inspect AI systems. Weak penalties for breaches are insufficient to deter deep-pocketed technology majors (Kuner, 2017).

For example, in India's DPDPA, penalties are bound and set by the government-constituted Data Protection Board, which is not structurally independent. In the United States, FTC action is reactive and limited in resources (Government of India, 2023). On the contrary, under the GDPR, fines up to 4% of a firm's turnover are allowed per year, but enforcement has varied across EU Member States (European Data Protection Board, 2023). Weak enforcement tools encourage developers and platforms to devalue safety and transparency in designing AI. In the absence of credible threats of legal repercussions, compliance is a matter of choice, not necessity.

### 11.7.7 Lack of direct obligations on AI developers for disinformation

Most data protection managements target controllers and processors. AI model trainers and developers, the pivotal players in the misinformation chain, are not directly subject to legal requirements in most jurisdictions. In the example of a generative AI model that generates disinformation health advice, it will not be liable under existing systems unless it manifestly breaches personal data standards (Malgieri, 2019).

This lack of responsibility is a regulatory blind spot. Developers can plead that once models are deployed and trained, users or platforms are responsible. Without a separate line of accountability, redress for those affected by AI-created misinformation is yet to happen swiftly.

### 11.7.8 Individual rights vs technological innovation

Both technological innovation and privacy of individuals are important. But technology cannot be promoted at the cost of immunity from accountability. Some arguments which portray those stringent and excessive rules against AI discourage innovation and startups represent a wrong notion. Such critics underestimate the importance of privacy rights of the individuals. These rights are inalienable rights and cannot be compromised at the cost of innovation. Both privacy and innovation need to be handled carefully without giving much precedence or priority against each other (Floridi et al., 2018).

The approach should be balanced to ensure checks and balances. The unchecked AI-based technology invites various challenges of deepfakes and misinformation that further escalates personalized rights of the individuals (D'Acquisto, 2024). The regulatory framework is the only solution that may balance the individual rights and technological innovations rights. The tech industry businesses have to follow and adhere to the regulations for a smooth governance model (Wachter & Mittelstadt, 2022).

## 11.8 CONCLUSION

Although the technological innovation has provided new opportunities through creative generative AI models to make life easy in various sectors, it has faced major governance challenges of privacy breaches of individuals. The study explored the failure of the existing laws in tackling gray areas of the AI in violating personal autonomy of the individuals. This chapter examined the complex issues of generative AI and the possible solution to ensure rights be protected through legal mechanism.

Developers and platforms are, nevertheless, shielded by fragmented legal obligations and outdated intermediary liability protections. Enforcement authorities, meanwhile, lack the financial, legal, and technical resources to investigate or prosecute harm effectively or in a timely manner. These are compounded by accelerating development rates of AI. It is now possible to generate voices, faces, and writing styles with shocking accuracy. This technology, while occasionally harmless in motivation, is readily exploitable for falsehood, trickery, and political manipulation. From election deepfakes to health disinformation generated by AI, the harms are now no longer hypothetical but are real, measurable, and growing. The slow legal regulations to address these harms create a perilous vacuum, under which the technological frontier can expand unchecked.

To understand the urgency of privacy data-related issues, several legal changes are required. First, data protection legislation must be updated to encompass AI-generated data and impose clear obligations on developers, deployers, and platforms. Second, international cooperation is essential because generative AI is transboundary in nature, and countries must engage in treaty negotiation, cross-border enforcement treaties, and standardization through international organizations like the UN, OECD, or G20. Third, there must be both technological and legal instruments in the hands of the regulators to conduct algorithmic checks, impose transparency requirements, and protect citizens rights.

Further, a policy of new digital rights is called for. These might include the right to explanation of algorithms, the right to object to machine-made representations, and the right not to be synthetically impersonated. Such rights would extend beyond informational privacy to promote cognitive liberty, reputational dignity, and democratic participation.

The future of generative AI will be shaped not only by what is possible but by what the society makes possible. Legal frameworks have a moral and constitutional responsibility to step in where technology erodes human rights, distorts public debate, or compromises the texture of democratic self-governance. Data protection laws, therefore, must evolve from a passive, retrospective shield to a proactive, anticipatory tool of justice.

In conclusion, the real threat posed by generative AI misinformation lies not merely in the sophistication of the technology but in the apathy of law.

If left unchecked, these systems will erode the very foundations of informed consent, autonomous identity, and truth itself. To avert this dystopian trajectory, lawmakers must act with urgency, wisdom, and resolve, crafting a data protection standard that not only responds to present challenges but also anticipates those on the horizon.

## REFERENCES

Ashutosh. (2024). Cross-border data flows and international law: Navigating jurisdictional complexities in the digital age. *Indian Journal of Law*, 2(1), 15–23. https://doi.org/10.36676/ijl.v2.i1.03.

Associated Press. (2024, February 1). Fake Biden robocall being investigated in New Hampshire. *AP News*. https://apnews.com/article/new-hampshire-primary-biden-ai-deepfake-robocall-f3469ceb6dd613079092287994663db5.

Barari, S., Lucas, C., & Munger, K. (2025). Political deepfakes are as credible as other fake media and (sometimes) real media. *Journal of Politics,* 87(2), 510–526. https://www.journals.uchicago.edu/doi/10.1086/732990.

Bincoletto, G. (2021). Data protection by design: From privacy by design to Article 25 of the GDPR. In *[Book Title Unknown]* (Chapter 2). Nomos Verlagsgesellschaft. https://www.nomos-elibrary.de/de/10.5771/9783748929895-37/chapter-2-data-protection-by-design-from-privacy-by-design-to-article-25-of-the-gdpr?page=1.

Brewster, T. (2021, October 14). Huge bank fraud uses deep-fake voice tech to steal millions. Forbes. https://www.forbes.com/sites/thomasbrewster/2021/10/14/huge-bank-fraud-uses-deep-fake-voice-tech-to-steal-millions/.

Bundesverfassungsgericht (Federal Constitutional Court). (1983, December 15). Census Act Case (Volkszählungsurteil), 65 BVerfGE 1. https://www.bundesverfassungsgericht.de/SharedDocs/Entscheidungen/EN/1983/12/rs19831215_1bvr020983en.html.

Cadwalladr, C., & Graham-Harrison, E. (2018, March 17). Revealed: 50 million Facebook profiles harvested for Cambridge Analytica in major data breach. *The Guardian*. https://www.theguardian.com/news/2018/mar/17/cambridge-analytica-facebook-influence-us-election.

California Legislature. (2018). California Civil Code §§ 1798.100–1798.199 (California Consumer Privacy Act of 2018, as amended by the California Privacy Rights Act of 2020). https://law.justia.com/codes/california/2018/code-civ/division-3/part-4/title-1.81.5/section-1798.100/.

Chesney, R., & Citron, D. (2019). Deep fakes: A looming challenge for privacy, democracy, and national security. *California Law Review,* 107(6), 1753–1819. https://www.californialawreview.org/print/deep-fakes-a-looming-challenge-for-privacy-democracy-and-national-security.

Citron, D. K. (2018). Extremist speech, compelled conformity, and censorship creep. *Notre Dame Law Review,* 93, 1035–1060. https://scholarship.law.nd.edu/ndlr/vol93/iss3/3/.

Cybersecurity and Infrastructure Security Agency. (2023, October). AI misinformation threat landscape assessment. U.S. Department of Homeland Security.

D'Acquisto, G. (2024). Synthetic data and data protection laws. *Journal of Data Protection & Privacy,* 6(3), 227–239. https://doi.org/10.69554/HFOS8421.

European Commission. (2021, April 21). Proposal for a regulation laying down harmonised rules on artificial intelligence (Artificial Intelligence Act) (COM(2021) 206 final). https://eur-lex.europa.eu/legal-content/EN/TXT/?uri=CELEX%3A52021PC0206.

European Data Protection Board, & European Data Protection Supervisor. (2021, June 18). Joint Opinion 5/2021 on the proposal for a Regulation of the European Parliament and of the Council laying down harmonised rules on artificial intelligence (Artificial Intelligence Act). https://www.edpb.europa.eu/our-work-tools/our-documents/edpbedps-joint-opinion/edpb-edps-joint-opinion-52021-proposal_en.

European Data Protection Board. (2018, May 25). Guidelines on automated individual decision-making and profiling for the purposes of Regulation 2016/679. European Data Protection Board.

European Data Protection Board. (2019, November). Guidelines 4/2019 on article 25 data protection by design and by default. European Data Protection Board. https://www.edpb.europa.eu/our-work-tools/our-documents/guidelines/guidelines-42019-article-25-data-protection-design-and_en.

European Data Protection Board. (2023, April 17). EDPB annual report 2022. European Data Protection Board. https://www.edpb.europa.eu/our-work-tools/our-documents/annual-report/edpb-annual-report-2022_en.

European Parliament, & Council of the European Union. (2016). Regulation (EU) 2016/679 of the European Parliament and of the Council of 27 April 2016 on the protection of natural persons with regard to the processing of personal data and on the free movement of such data (General Data Protection Regulation). *Official Journal of the European Union, L*119, 1–88. https://data.europa.eu/eli/reg/2016/679/oj.

European Parliament. (2017, January 12). Civil law rules on robotics (2015/2103(INL)). https://www.europarl.europa.eu/doceo/document/TA-8-2017-0051_EN.html.

European Union. (2000). Charter of fundamental rights of the European Union, 2000 O.J. (C364) 1. https://eur-lex.europa.eu/legal-content/EN/TXT/?uri=CELEX%3A12012P%2FTXT.

European Union. (2016a). General Data Protection Regulation (EU) 2016/679, Chapter V: Transfers of personal data to third countries or international organisations. *Official Journal of the European Union*, *L*119. https://gdpr-info.eu/chapter-5/#:~:text=Protection%20Regulation%20(GDPR)-,Transfers%20of%20personal%20data%20to%20third%20countries%20or%20international%20organisations,the%20protection%20of%20personal%20data.

European Union. (2016b). Regulation (EU) 2016/679 of the European Parliament and of the Council of 27 April 2016 on the protection of natural persons with regard to the processing of personal data and on the free movement of such data (General Data Protection Regulation). *Official Journal of the European Union, L*119, 1–88. https://eur-lex.europa.eu/eli/reg/2016/679/oj.

European Union. (2016c). Regulation (EU) 2016/679 of the European Parliament and of the Council of 27 April 2016 on the protection of natural persons with regard to the processing of personal data and on the free movement of such data (General Data Protection Regulation), Article 5,6,9 & 22. *Official Journal of the European Union*. https://eur-lex.europa.eu/eli/reg/2016/679/oj.

European Union. (2016d). Regulation (EU) 2016/679 of the European Parliament and of the Council of 27 April 2016 on the protection of natural persons with regard to the processing of personal data and on the free movement of such data (General Data Protection Regulation), Article 25. *Official Journal of the European Union*, L119, 1–88. https://eur-lex.europa.eu/eli/reg/2016/679/oj.

European Union. (2024). Artificial intelligence act (regulation (EU) 2024/1689). *Official Journal of the European Union*, *L*169/1. Entered into force August 1, 2024. https://eur-lex.europa.eu/eli/reg/2024/1689/oj.

European Union. (2024). Artificial intelligence act. *Official Journal of the European Union,* 1–44. https://eur-lex.europa.eu/legal-content/EN/TXT/PDF/?uri=OJ:L_202401689

Federal Trade Commission. (2023, April). The lure of generative AI and the risk of misinformation. https://www.ftc.gov/policy/advocacy-research/tech-at-ftc/2023/06/generative-ai-raises-competition-concerns.

Federal Trade Commission. (2023, August 22). FTC action stops business opportunity scheme that promised its AI-boosted tools would power high earnings through online stores [Press release]. FTC. https://www.ftc.gov/news-events/news/press-releases/2023/08/ftc-action-stops-business-opportunity-scheme-promised-its-ai-boosted-tools-would-power-high-earnings.

Fenwick, M., Siems, M., & Vermeulen, E. P. M. (2009). The shifting meaning of legal certainty in comparative and transnational law. *Brooklyn Journal of International Law,* 34(1), 107–151. https://kyushu-u.elsevierpure.com/en/publications/the-shifting-meaning-of-legal-certainty-in-comparative-and-transn.

Floridi, L., & Cowls, J. (2019). A unified framework of five principles for AI in society. *Harvard Data Science Review,* 1(1). https://doi.org/10.1162/99608f92.8cd550d1.

Floridi, L., Cowls, J., Beltrametti, M., Chatila, R., Chazerand, P., Dignum, V., ... Vayena, E. (2018). AI4People—An ethical framework for a good AI society: Opportunities, risks, principles, and recommendations. *Minds and Machines,* 28(4), 689–707. https://doi.org/10.1007/s11023-018-9482-5.

Goodfellow, I. J., Pouget-Abadie, J., Mirza, M., Xu, B., Warde-Farley, D., Ozair, S., Courville, A., & Bengio, Y. (2014). Generative adversarial nets. In *Advances in Neural Information Processing Systems,* 27 (NIPS 2014). Montréal, Quebec, Canada. https://papers.nips.cc/paper_files/paper/2014

Government of India. (2023). Digital personal data protection act, 2023, Chapter VI & VIII. Ministry of Law and Justice.

Guimarães Moraes, T. (2024). Ethical AI regulatory sandboxes: Insights from cyberspace regulation and Internet governance. In *Proceedings of the 2nd International Symposium on Trustworthy Autonomous Systems* (pp. 1–10). ACM, University of Texas USA. https://doi.org/10.1145/3686038.3686049.

Habermas, J. (1989). *The Structural Transformation of the Public Sphere.* Cambridge: Polity Press.

Huq, A. Z. (2013). Removal as a political question. *Stanford Law Review,* 65(1), 1–56.

Jobin, A., Ienca, M., & Vayena, E. (2019). The global landscape of AI ethics guidelines. *Nature Machine Intelligence,* 1(9), 389–399. https://doi.org/10.1038/s42256-019-0088-2.

Kreps, S., & Kriner, D. (2023). How AI threatens democracy. *Journal of Democracy,* 34(4), 122–131. https://www.journalofdemocracy.org/articles/how-ai-threatens-democracy/.

Kuner, C. (2017). Reality and illusion in EU data transfer regulation post Schrems. *German Law Journal,* 18(4), 881–918. https://doi.org/10.1017/S2071832200022197.

Kuner, C., Svantesson, D. J. B., Cate, F. H., Lynskey, O., & Millard, C. (2017). Machine learning with personal data: Is data protection law smart enough to meet the challenge? *International Data Privacy Law,* 7(1), 1–2. https://doi.org/10.1093/idpl/ipx003.

Malgieri, G. (2019). Automated decision-making in the EU Member States: The right to explanation and other "suitable safeguards" in the national legislations. *Computer Law & Security Review,* 35(5), Article 105327. https://doi.org/10.1016/j.clsr.2019.05.002.

Mill, J. S. (1859). *On Liberty*. London: John W. Parker and Son.

Mittelstadt, B. D., Allo, P., Taddeo, M., Wachter, S., & Floridi, L. (2016). The ethics of algorithms: Mapping the debate. *Big Data & Society,* 3(2), 1–21. https://doi.org/10.1177/2053951716679679.

Narayanan, A., & Shmatikov, V. (2008). Robust de-anonymization of large sparse datasets. In *2008 IEEE Symposium on Security and Privacy (sp 2008)*, Oakland, CA (pp. 111–125). IEEE. https://doi.org/10.1109/SP.2008.33.

National Institute of Standards and Technology. (2023, January 26). AI risk management framework (AI RMF 1.0). NIST. https://www.nist.gov/itl/ai-risk-management-framework.

O'Sullivan, S. (2023). AI, regulation and the speed of technological development. *Computer Law & Security Review*, 48, Article 105843.

OECD. (2019). Recommendation of the council on artificial intelligence. OECD Publishing. https://legalinstruments.oecd.org/en/instruments/oecd-legal-0449.

Organisation for Economic Co-operation and Development (OECD). (2019). Recommendation of the council on artificial intelligence. OECD Publishing.

Pagallo, U. (2013). *The Laws of Robots: Crimes, Contracts, and Torts*. Dordrecht: Springer.

Pasquale, F. (2015). *The Black Box Society: The Secret Algorithms That Control Money and Information*. Cambridge, MA: Harvard University Press.

Pavis, M. (2021). Rebalancing our regulatory response to deepfakes with performers' rights. *Convergence,* 27(4), 974–998. https://doi.org/10.1177/13548565211033418.

Prosser, W. L. (1960). Privacy. *California Law Review,* 48(3), 383–423. https://doi.org/10.15779/Z383J3C.

Rawls, J. (1971). *A Theory of Justice*. Cambridge, MA: Harvard University Press.

Roberts, H., Hine, E., Taddeo, M., & Floridi, L. (2024). Global AI governance: Barriers and pathways forward. *International Affairs,* 100(3), 1275–1286. https://doi.org/10.1093/ia/iiae073.

Sahoo, N. (2021). Right to be forgotten in India: Case law and emerging trends. ORF Issue Brief. Observer Research Foundation.

Schwartz, P. M. (2004). Property, privacy, and personal data. *Harvard Law Review,* 117(7), 2056–2128. https://doi.org/10.2307/4093335.

Schwartz, P. M., & Solove, D. J. (2011). The PII problem: Privacy and a new concept of personally identifiable information. *New York University Law Review,* 86(6), 1814–1894. https://www.scirp.org/reference/referencespapers?referenceid=3451510.

Solove, D. J. (2002). Conceptualizing privacy. *California Law Review,* 90(4), 1087–1155. https://doi.org/10.2307/3481326.
Solove, D. J. (2008). Understanding privacy. Harvard University Press. https://scholarship.law.gwu.edu/faculty_publications/922/.
Supreme Court of India. (2017). Justice K.S. Puttaswamy (Retd.) v. Union of India, (2017) 10 SCC 1.
Tufekci, Z. (2014). Engineering the public: Big data, surveillance, and computational politics. *First Monday,* 19(7). https://doi.org/10.5210/fm.v19i7.4901.
U.S. Congress House Committee on Commerce, Science, and Transportation. (2023). Algorithmic accountability act of 2023, H.R. 6580, 118th Cong. (2023). Retrieved from Congress.gov. https://www.congress.gov/bill/118th-congress/senate-bill/2892/text.
United States Code. (1996). Communications decency act, 47 U.S.C. § 230.
Veale, M., & Edwards, L. (2018). Clarity, surprises, and further questions in the Article 29 Working Party draft guidance on automated decision-making and profiling. *Computer Law and Security Review*, 34(2), 398–404. https://doi.org/10.1016/j.clsr.2017.12.002.
Virginia General Assembly. (n.d.). Virginia code § 24.2–956. Code of Virginia. https://law.lis.virginia.gov/vacode/title24.2/chapter9.5/section24.2-956/.
Wachter, S., & Mittelstadt, B. (2022). A right to reasonable inferences: Re-thinking data protection law in the age of big data and AI. *Columbia Business Law Review*, *2022*(3), 687–757. https://doi.org/10.7916/cblr.v2019i2.3424.
Wardle, C. (2017, February 16). Fake news. It's complicated. First Draft News. https://firstdraftnews.org/articles/fake-news-complicated/.
Wardle, C., & Derakhshan, H. (2017, September). Information disorder: Toward an interdisciplinary framework for research and policymaking. Council of Europe. https://rm.coe.int/information-disorder-toward-an-interdisciplinary-framework-for-researc/168076277c.
Westin, A. F. (1967). Privacy and Freedom. Atheneum.
Xie, Z. (2025). Who is spreading AI-generated health rumors? A study on the association between AIGC interaction types and the willingness to share health rumors. *Social Media + Society,* 11(1). https://doi.org/10.1177/20563051251323391.
Zuboff, S. (2019). *The Age of Surveillance Capitalism*. PublicAffairs.

Chapter 12

# Emerging trends and challenges in generative artificial intelligence (GenAI) and misinformation

## HR perspectives

*Revati Ramrao Rautrao, A. V. Senthil Kumar, and T.S. Sanjayan*

### 12.1 INTRODUCTION

Generative Artificial Intelligence (GenAI) is rapidly redefining Human Resource Management (HRM), transforming core functions such as talent acquisition, performance evaluation, and employee engagement through automation, personalisation, and predictive analytics (McKinsey, as cited in AIHR, 2024). While GenAI provides significant operating benefits, it also introduces complex risks, especially the spread of misunderstandings, including fabricated credentials, biased production, and misleading communication (Caproro et al., 2023).

This chapter examines the double effect of GenAI on HR ecosystems: its ability to increase efficiency and innovation and its tendency to generate misinformation that reduces organisational beliefs, culture, and decision-making. GenAI competition systems are used quickly for automated resume screening, adaptive education modules, and strategic workforce schemes. Nevertheless, these structures have ethical and symbolic risks and might result in hallucinations, deepfake communication, and deviant internal signals (Giralt Hernández, 2024; Sadeghi, 2024). A planned reaction to the mixing of GenAI is needed for HR capabilities, one that strikes a balance between technical development and legal, moral, and ethical requirements. In addition to being pragmatic directors, HR experts need to negotiate this evolving environment as moral advocates for using AI. They should try this to align the distribution of GenAI with the corporation's ideas of inclusivity, integrity, and reputation. Although GenAI promises to enhance HR, its implementation additionally creates more tough moral and realistic issues. The following sections explain how these techniques are currently used in HR functions—and where their risk begins on the surface. This chapter is directed by three theoretical lenses: social technology theory, organisational justice theory, and technology acceptance models, which are used to explain the GenAI's effect on HR practice, moral risks, and user engagement.

 DOI: 10.1201/9781003626558-16

## 12.1.1 Background

### *12.1.1.1 Context of generative artificial intelligence (GenAI) in organisations*

GenAI has emerged as a transformational force in organisational contexts, especially in knowledge-intensive areas. The ability to generate lessons, images, and code has enabled new forms of automation, creativity, and decision support in business functions. In HRM, GenAI applications are used to streamline rapid recruitment, to adapt employees' learning, and optimise the workforce (AIHR, 2025). Xu et al. (2025) argue that GenAI reduces information overload by increasing workers' capacity to coordinate complex problems and manage cognitive requirements. This change is particularly relevant in a dynamic environment where rapid decision-making and adaptive learning are needed. However, integration of GenAI also introduces new weaknesses. Trained systems on biased or incomplete data can provide misleading outputs, strengthen existing differences, and erode confidence in organisational processes (Budhwar, 2023). According to TBS Education (n.d.), GenAI Tools such as ChatGPT actively explore their potential to personalise HR workflows, yet their deployment requires careful oversight to prevent unintended consequences. The dual nature of GenAI—as both an enabler and a disruptor—necessitates a strategic approach that aligns technological capabilities with ethical governance and stakeholder accountability.

The discovery of mutual activity between GenAIand misinformation in HRM is important because of the three interconnecting imperatives. First, GenAI has moved the extent of automation, and normal materials have been transferred to production. Now, AI system candidates can add a qualification, create an internal memorandum, or mimic a control response without human monitoring and blur the line between valid and artificial. Stories with this functionality are much more likely to be incomplete. Second, social and cultural ecosystems clearly comprise HR domains. The administrative centre, company area of expertise, and worker morale are all highlighted within the HR selections, which move beyond easy capability. Inaccurate records diminish those sensitive social garments: a manufacturing unit's launch may be a raga with an extensive variety of options, a mistake can cause operational uncertainty, and a mistake can harm the recognised capital. The function of HR in these conditions goes beyond coping with self-perception. Third, HR alternates between fragmented, misleading statements and AI. Although innovation is recommended, the process of detection and backlog is due by ethical recommendations, particularly regarding inaccurate information. This divergence places HR in an unsure area that creates stability among technological adoption and aspects of cultural, ethical, and legal security.

#### *12.1.1.2 Implications for HRM*

Regular synthetic intelligence (GenAI) fulfils both the latent and transformational ability. Quick Talent Collection, a dynamic training tool, and vacation spot competencies in motion planning are all effective for HR leaders from a strategic perspective. This innovation can enable HR to work with extra self-assurance and recognition on better-order features, with improvement from a strategic group of workers. However, misinformation-stimulated disruption outcomes are non-negligible. Produced or biased recruitment selections can lessen the organisational range and fairness, and inner verbal exchange can lead to misinterpretation of politics; deepfakes may be made weapons for the loss of popularity. In addition, employees are threatened by uncontrolled automation, extended concerns, links, or a demon setting. Therefore, HR should play an expanded role of an architect and a mentor for AI trust in the organisation. The integration of GenAI provides transformational ability to HR but carries latent dangers. Strategically, HRMs regularly benefit from employing these techniques, distributing adaptive teaching plans, and estimating the change in the workforce. HRMs can be free of regular tasks and can make more efforts for professional strategic workforce development, organisational welfare, and human-focused policy design. Conversely, misinformation-related disorders are far from trivial. As highlighted by Chaturvedi and Chaturvedi (2025), biased recruitment decisions driven by GenAI systems can undermine justice and inclusion. Furthermore, misleading internal communication and deepfake manipulations pose reputational risks that erode stakeholder trust and organisational cohesion. In addition, heavy dependence on automation can distinguish between employees and decrease stress, anxiety, and organisational harmony. These unexpected results require the useof HR-only not only as an advanced operator of the AI tool but also as a moral protector of AI-operated changes.

An integrative theoretical lens helps illuminate how GenAI and misinformation interact within HR contexts. Three foundational perspectives are particularly relevant, and an integrated theoretical lens helps illuminate how to interact with genes and misinformation in HR references. These three approaches are particularly relevant:

a. **Sociotechnical systems theory**: This framework emphasises the interaction between technical systems and social structures. HRA scoping review (2024). The effect on HR activities has been emphasised, including how automation, task design, and social relationships develop as well as technology, and confirms that effective HRMs should be responsible for both technical and human dimensions, PMC. This perspective emphasises that technology implementation must be considered in the social environment. GenAI does not work in a vacuum; its effects take the form of organisational norms, human interactions, and cultural expectations. Strategic, human-centric role: HRMs should coordinate technology design with human ideas and confirm that technical interventions require strong sociotechnical integration to gain trust and efficiency.

b. **Organisational justice theory:** The perceptions of justice—the popular, procedural, and interpersonal—are central to how employees assess organisational functions. Robert et al. (2020) advocated the design of an AI system that maintains justice in these dimensions. Procedural justice requires openness about how genius-operated HR decisions are made; interactional justice requires respectful and real human communication; and distribution ensures comparable results in justice groups. Regardless of these dimensions, GenAI reduces the workforce and organisational validity. Inherent in the perception of justice—hereditary, procedural, and intravenous—this theory provides an objective evaluation of how genius systems affect employees' perceptions of justice. Robert et al. (2020) proposed a design agenda focusing on this type of justice proposal and suggested that AI must be structured to support the results only in HR decisions and arXiv.

c. **Technology Acceptance Models (TAM & UTAUT):** Recent HR studies with integrated principles for the acceptance and use of technology models (TAM) and technology (UTAUT). Prasad et al. found that "utility" and "ease of use" affect the confidence of employees and connection to the genius tool and serve organisational involvement with the confidence and intermediary of the results. This indicates that the moral deployment of GenAI shoul emphasise transparent, dependable, and user-friendly design. TAM and their modern expansion, out power, specify that the experience of using a technology and its ease of use affect its acceptance; confidence acts as an important intermediary. Prasad and De demonstrated that in HR references, the notion of employees in genius units affects organisational involvement, employee involvement, and performance, while confidence partially conveys these conditions. Similarly, Korzyński et al. noted that top management supports Foster Trust, which in turn forms actual GenAI adoption. This insight confirms that for a moral, effective GenAI deployment in HR, the system should be presented within a transparent, dependable, and supportive leadership structure.

This chapter presents a distinct contribution by integrating the emerging issue of GenAI-driven misinformation with foundational HR theory. Unlike previous studies that emphasised automation and efficiency gains (Tambe et al., 2019), this chapter foregrounds the socioemotional and cultural fallout of AI-mediated misinformation in HR contexts. By leveraging the sociotechnical systems theory, we illuminate how human and technological subsystems must be co-designed to mitigate misinformation risks. By incorporating organisational justice theory, we underscore the necessity of upholding procedural, distributive, and interactional fairness when deploying GenAI in sensitive HR scenarios (Robert et al., 2020). Further, drawing on technology acceptance models and contemporary trust frameworks, we argue that a user's acceptance of GenAItools is contingent not only on usability but also on perceived fairness and transparency (McGrath et al., 2024). Budhwar (2023) underscores the transformative potential of GenAI in HRM, particularly in redefining recruitment and learning through conversational agents like ChatGPT, thereby opening new avenues for strategic workforce engagement.

Our primary innovations are threefold:

1. **Diagnostic precision:** Mapping specific misinformation threats across HR domains such as recruitment, learning, and internal communication.
2. **Theoretical synthesis:** Combining sociotechnical, justice, and trust literatures to advance actionable understanding.
3. **Strategic constructs:** Everything from the Human-in-loop protocol to the virtual literary training protocol emphasises Chowk's validation. All things taken into consideration, their contributions allowed HR leaders to be flexible and ethical.

### 12.1.2 Methodology

This bankruptcy adopts an ideological synthesis feature, which integrates interdisciplinary concepts and empirical insights, and examines the intersection of established synthetic intelligence (GenAI) in HRM. Instead of relying on primary records series, evaluation is interested in developing a thematic shape from colleague assessment literature, political inclination, and industry evaluations that reflect innovative traits, new dangers, and uncertainties. The observation is in three theoretical lenses: the sociotechnical machine theory, which examines the development of human actors in AI technology and organizational contexts; organisational law idea, which shows the moral assessment of talent applications in relation to justice, openness, and procedural integrity; and era popularity fashions (TAME and UTAUT), which lead the know-how of the person's involvement, self-warranty, and resistance to AI-driven systems. This technique is designed for HR doctors, decision-makers, and researchers via mapping the moral, operational, and regulatory dimensions of genes and skills. Finally, this chapter continues with a human-focused and governance-oriented perspective when AI adoption is responsible for HRM. The conceptual analysis performs a vital function in growing moral, operational, and control troubles before mass studies are feasible. This chapter aims to perceive predominant dangers and strategic reactions and create a future study structure.

### 12.1.3 Structure of this chapter

The remaining part of this chapter is structured as follows: Section 12.2 presents a synthesis of recent literature and ideological reviews related to the integration of GenAI into HRM with an emphasis on moral and operating dimensions. Section 12.3 emphasises the biggest challenges related to AI-operated misinformation in the HR system, including data construction, algorithmic level, and lack of transparency. Section 12.4 discusses the existing mitigation strategies adopted by HR subjects and institutions to address these risks, such as the digital literature initiative, the moral steering model, and the verification protocol. Section 12.5 introduces future paths for GenAI, focusing on strategic adaptation to ESG goals, human-in-the-loop collaboration, and active risk management. Section 12.6 examines the regulatory structure and moral imperative for the adoption of GenAI, with references. Finally, Section 12.7 concludes the chapter by summarising the most important insights and by providing recommendations to researchers, doctors, and policy-makers, who navigated the landscape developed in HRM.

## 12.2 RELATED WORK

Integration of GenAI into HRM has attracted attention from researchers in recruitment automation, private teaching, and destination analysis. Recent research shows GenAI's competent operational abilities, including emotion-based performance assessment, screening of applicants, and an adaptive training platform. The concerns of algorithm changes, privacy, and ability to dissolve are problems with morality and governance, which have been addresses by a parallel chapter on the supply of AI in HR. Studies provide an integrated structure related to both revolutionary capacity and systemic errors in GenAI, although current research provides practical information on the technical and management aspects of body technology. This is the most important functionality, which combines multidisciplinary methods and provides a management-oriented approach to reduce misinformation and encourage moral AI integration in HR, produced by this insolvency. GenAI quickly redefines HR by enabling advanced automation, individual employees' experiences, and data-driven decisions. The recruitment methods are quickly promoted with the aid of restarting the AI purchased by means of screening, candidate matching, and forecasting overall performance evaluation, lowering bias, and accelerating the ticketing process. In learning and development, GenAI sequential training modules, adaptive skill assessment, interactive simulation, and employee involvement increaseemployee involvement and discipline. Performance management benefits from the real-time reaction of AI facilities, future storage models, and dynamic target determination, combined with organisational goals. In addition, the GenAI landscape supports the work plan through modelling and strategic resource allocation. These applications are mature in terms of moral thoughts and openness and not only optimise operational efficiency but also cultivate a more legitimate, tight, and data-informed HR ecosystem.

### 12.2.1 Recruitment and talent acquisition

GenAI prepares recruitment by competing by enabling intelligent resume screening, automated candidate interaction, and predictive job-fit analysis. These technologies provide sufficient efficiency benefits, restructuring the hiring process and helping applicants improve the adjustment between profiles and role requirements. For example, Khan et al. (2024) show that candidates can reduce the GenAI system from time to time by increasing matching accuracy. The platforms that use AI-operated chatbots and analytics have shown the ability to cut 30%–50% the recruitment cycle so that the HR teams can focus on strategic engagement from administrative functions. However, there is a rapid adoption of moral and operational challenges. Gartner (2024) warns that opaque algorithm processes can strengthen systemic prejudice, especially when training data reflect historical inequalities. Evidence indicates that GenAI screening equipment may inadvertently take advantage of male candidates or reintroduce demographic and educational prejudices (Chaturvedi and Chaturvedi, 2025). These risks include algorithmic designs, transparent decision-making, and human inspection requirements. Organisations should apply the verification protocol and the bias audit to ensure that the pursuit of efficiency does not conflict with justice, equity, or stakeholder trust.

### 12.2.2 Learning, development, and employee engagement

The GenAI forms organisational learning by activating adaptive training modules, personal skill assessment, and dynamic career development paths. Platforms such as 365s show how GenAI can identify personal competences, match employees with relevant learning opportunities, and promote continuous professional growth in the inclusive, data-mechanised ecosystem. In parallel, GenAI-operated communication equipment provides real-time guidance and supports employee involvement by supporting the hybrid work environment. Swift and Business Insider highlight how AI-operated agents can provide personal information, policy updates, and interactive support, increasing the continuity and accountability of distributed teams. Initial evaluation suggests that a GenAI-competent teaching system can significantly improve staff involvement and organisational adjustment. However, cultural sensitivity and emotional shading borders affect these advantages. ResearchGate cautions that while the AI interface offers balance and scalability, it cannot have the specified consciousness and empathy that unavoidably affect human symptoms. Furthermore, the education risk of the GenAI system may additionally exacerbate bias in schooling materials or fail to adapt to various learning styles if planned safety precautions aren't taken. To reduce these dangers, groups must combine algorithm privatisation with human monitoring and make certain that the schooling materials are thorough, ethical, and pertinent. The authority of university students, reference verification, and continuous response loops are vital for flexible, proficient, and entire ecosystems that support both personal growth and organisational integrity.

### 12.2.3 Workforce planning and HR decision support

Destiny Forecast, Dynamic Performance Management, and real-time emotional assessment are made viable through GenAI, which revolutionises the operating group. Using sophisticated generation and herb language remedies, **Management Resources** (MR) can determine management pipelines, degree of professional understanding, and align human capital techniques with long-term enterprise goals (Tambe et al., 2019). The market alerts of GenAI people are incorporated with externally difficult obligations, unlike conventional opportunity assistance systems. In this manner, gaps, threats, and objective individuals must be addressed so that one can provide the possibility to broaden granular insight into the needs. We ought to make use of flashy, panorama-based modelling and spread the word about this HR for the use of the reactive approach. For example, the fulfilment strategy now includes organisational network mapping and behavioural evaluation, in addition to the previous fashionable performance metrics, to guarantee continuity and flexibility. This shift has its theoretical roots in institutional entrepreneurship, which highlights how businesses employ era innovation to strike a balance between validity and flexibility. When applied to HR, this standpoint implies that GenAI now not only robotically produces employee dynamics but additionally makes it feasible for them to construct adaptive institutional reflexes. Human collaboration is required. Distribution must comply with moral evaluation and pertinent judgments, even though GenAI can streamline planning methods. HR leaders need to make certain that the set of rules tips are legitimate for organisational values and strategic choices. Working on the motion plan isn't always best as a hit technical exercise, but also a strategic collaboration that complements agility, obligation, and schooling for destiny disruptions by approaching GenAI in a sociotechnical way. Despite developments in recruiting, getting to know, and motion strategies, there are still some drawbacks. The next phase evaluates how the popularity of facts—which incorporates opaque production, algorithm bias, and the integrity of choice-making techniques—is misinterpreted.

## 12.3 MISINFORMATION RISKS: ETHICAL FAULT LINES IN HR

GenAI can revolutionise HR physical games, but it additionally highlights significant concerns associated with fake information. It is not unusual for massive language fashions that prioritise the GenAI system to provide outputs that seem coherent but are clearly misguided or inappropriate; this phenomenon is known as "hallucinations" (Gartner, 2025). Such inaccuracy can be visible in a few vital regions of HRM, along with workforce evaluation, political communication, and screening, which may additionally result in bad decisions and ethical quandaries. AI candidates, as an example, should have credentials or overstated performance on their profiles, which diminishes fairness and inclusivity in hiring techniques. (Weidinger et al., 2021). Likewise, GenAI-generated inner memoranda or onboarding materials might also disseminate ambiguous or fake statistics, which might undermine workers'acceptance and corporate transparency. The intricacy of GenAI actual-time verification makes it hard and leaves HR in an undefined position while assessing the validity of AI-generated documents, in contrast to traditional structures, wherein faults are found. GenAI can doubtlessly be complicated in the absence of human scrutiny. HR administrators may also struggle to guard the results or deal with worries while selections are made primarily based on the output of an opaque set of rules. In addition to elevating the threat of iconic harm, this uncertainty now attracts attention to legal investigative firms, mainly in courts with stringent anti-discrimination and statistics safety rules (Reuters, 2024). Organisations need to adopt a sturdy management framework that includes human-in-loop protocols, numerous layers of confirmation structures, and ongoing AI literature education, which will decrease these dangers. These safety precautions assure that GenAI enhances HR innovation without compromising ethical ideas or the legitimacy of the business enterprise. HR executives can utilise the casting approaches organised in Section 12.4 to guard against disinformation managed by GenAI.

### 12.3.1 Fabricated data and hallucinated content

One of the most published risks associated with GenAI in HR is the generation of manufactured or hallucinatory information, which appears to be admirable, but is wrong, misleading, or fully fabricated. These hallucinations can be displayed in various references for HR, including screening of re-starting, candidates such as profiles, and internal communication. Caproro et al. (2023) Note that the GenAI system, when indicated with incomplete or vague data, can produce synthetic credentials, exaggerated performance, or imaginary connections that distort recruitment decisions. This risk is expanded when the GenAI tool is used to generate the performance summary without boarding materials, policy documents, or without adequate human inspection. AIHR (2024) said that such materials, if not uncontrolled, can destroy confidence and compromise organisational openness.

A remarkable example was in 2024 when a multinational bank put out GenAI for internal communication. During the weeks, employees reported confusion due to the hallucination policy update—the implementation of a look-back and manual review protocol (Business Insider, 2025). This phenomenon outlines the recognised and operating risk by relying on GenAI without adequate monitoring. Unlike traditional errors, it is often difficult to detect hallucination results when mimicking the tone and structure of valid information, making them particularly dangerous in anenvironment with long hours. In addition, hallucinations can promote misinformation in digital platforms, especially when reused or recycled without confirmation. It creates a feedback loop where false information is reinforced, affecting potentially strategic decisions and assumptions about efforts. Gartner (2025) emphasises the requirements for the verification protocol at respective levels, which involvecross-reference with confirmed databases and a built-in human–lion mechanism to ensure material integrity. To reduce these risks, HR leaders must establish clear limits for genes, define verification items, and cultivate digital reading skills among employees. Considering GenAI as a co-pilot rather than the only decision-maker, organisations can use their abilities and protect against iconic and operational damage caused by fabricated materials.

### 12.3.2 GenAI in HR: promise and peril

GenAI spreads information or hallucinations in competent human resource processes and emphasises the status of systematic authentication and inspection. Unlike traditional HR equipment, generic models depend on potential logic and large-scale pattern recognition, giving results for relevant impurities. Organisations should hire multistage verification frameworks that integrate algorithm safety protections with human judgment if they want to acquire this. One approach is to enter the authenticity-checking set of rules, which compares the cloth produced with the aid of GenAI with a proven organisational database or legitimate outdoor assets. For example, AI-generated or altered resumes might be routinely checked against professional databases or qualifications to guarantee hiring rights. Like this, HR communications regarding GenAI should undergo a systematic assessment process where the leader or in-house officer confirms that the message is correct, clean, and compliant with regulatory norms. Organisational AI is a similarly widespread literacy improvement. HR ensures training as an improvement tool rather than an autonomous decision-maker to evaluate AI outputs, ask questions about deviations, and implement moral arguments. The oversight board, modelled according to company management structures, can make responsibility more institutional by monitoring oversight boards, AI payment, and audit decision results. Thus, resilient verification and overlooking mechanisms are not only corrective security measures but also, in GenAI-competent HR practice, basic columns to enter trust, openness, and moral flexibility.

### 12.3.3 Algorithmic bias and exclusion

Algorithmic bias and exclusion represent a critical concern in GenAI-enabled HR systems, particularly when these tools are used to evaluate candidates, recommend promotions, or personalise training pathways. Bias can emerge from the data used to train GenAI models, often reflecting historical inequalities and entrenched stereotypes (Gartner, 2024). In 2023, a leading global tech firm piloted GenAI-based screening tools to streamline its graduate recruitment process. Internal audits later revealed that the system disproportionately favoured candidates from elite institutions and underrepresented women in STEM roles. The bias was traced to historical training data that overemphasised prior hiring patterns, prompting the company to suspend the tool and initiate fairness recalibration protocols (Gartner, 2024). When left uncorrected, these biases can be amplified by the algorithm, resulting in discriminatory outcomes that undermine fairness and inclusion. For example, Chaturvedi and Chaturvedi (2025) report that GenAI-based screening tools may exclude male candidates or punish applicants from nonconventional educational backgrounds. Such prejudices do not always end; they can be built into language patterns, score algorithms, or decision-making limits that affect the marginalised groups. Wen et al. (2025). Furthermore, good-intense AI systems can also eliminate exclusion if justice calculations are clearly defined and monitored. Human inspection must be built into all stages of the AI life cycle—from model development to distribution and evaluation. Algorithms, by prioritising justice and transparency, can ensure that GenAI improves equity instead of increasing existing inequalities.

As shown in Table 12.1, while the GenAI apps offer benefits of skills, such as screening and adaptive teaching, they also introduce misunderstandings, including factory credentials and biased instructional material.

### 12.3.4 Transparency threats and organisational trust

Openness is a basic principle of moral HR exercises, but the genes often function as opaque "black box" models, making it difficult for HR to understand how decisions occur. This lack of interpretation creates serious challenges for organisational trust, especially in AI-driven production recruitment, results evaluation, or internal communication (Gartner, 2024). When employees and stakeholders are unable to detect the logic behind the

*Table 12.1* Generative artificial intelligence (GenAI) benefits vs. misinformation risks in HRM

| *Generative artificial intelligence (GenAI) application* | *Benefit* | *Misinformation risk* |
|---|---|---|
| Resume screening | Faster candidate shortlisting | Fabricated credentials or AI-generated false data |
| Adaptive learning | Personalised training experiences | Misleading content or biased instructional material |
| Workforce analytics | Predictive modelling for talent needs | Skewed interpretations due to biased datasets |
| Internal communications | Automated drafting and dissemination | Ambiguous or misleading messaging |

GenAI-created recommendations, trust in HR processes can be destroyed. For example, if a candidate is rejected based on algorithmic scoring without a clear argument, the decision can be considered arbitrary or biased. Cornell Law School emphasises that such ambiguity can expose organisations to legal investigation, especially under anti-discrimination and data transparency rules. In addition, repetition of vague claims about the risk of GenAI can reduce immediate moral governments without offering action-rich insights. Like the study by Simon at al. (2023), the discussion of logic and openness failed to solve deep institutional challenges that arose from AI adoption. HR leaders should be beyond the problem of accepting the problem and developing the framework, such as the underlying clarity, the stakeholder gauge in the audit unit, and the genius system and mustrestore and maintain organisational faith; openness must be maintained. This involves documentation of AI deposition routes, which allow the system to exaggerate, and how to clearly use the GenAI tool with employees. By organising technical distribution with moral clarity, HR can ensure that the genes are strong, withoutreducing social agreement between organisations and their people.

## 12.3.5 Workplace dynamics and dehumanisation

Artificial intelligence (GenAI) is used notably in HRM, which places an exceptional deal of stress on the state of affairs' effectiveness and obligation. Although those systems provide beneficial records, the mechanics behind their inherent alternatives are frequently difficult to understand, which results in the method (Reuters, 2024). HR experts may be capable of justifying themselves in these conditions by stating that an algorithm is associated with a specific undertaking idea, campaign results, or selections. Openness is destroyed by this lecturer's absence, which makes it difficult for stakeholders to protect findings, clear up conflicts, or adhere to painting regulations. When GenAI consists of statistical resources from behavioural evaluation to signs of the external exertions market, the issue gets even more complicated since the calculation method becomes even more complex. Employees sought proof of equity, contested human source picks, and had been denied probabilities; however, folks who rely upon professional, opaque algorithms may not have a defence. This lack of traceability makes criminal challenges more common in courts with anti-discrimination and hiring regulations. Furthermore, a lack of openness erodes consideration in organisational methods. Activists who believe that decisions are arbitrary or unclear are more likely to use AI or raise questions about institutional integrity. Tumbling these dangers, agencies should prioritise readability through explanatory models, robust documentation, and an obvious control shape. Without these safety features, typical GenAI is responsible for innovation coverage.

## 12.4 STRATEGIC MITIGATION: RECLAIMING CONTROL IN AI-AUGMENTED HR

To address the risk generated by misinformation, HR leaders should adopt active mitigating strategies connecting technical security measures with moral governance. These strategies should be used in recruitment, learning, and work planning processes to ensure that talent improves organisational results without compromising integrity or trust.

Verification protocol: One of the most effective approaches is the implementation of verification systems at numerous levels. These include cross-referencing GenAI output with a valid database, the manual revision of the candidate profiles, and using metadata tracking to certify material sources. As mentioned in Table 12.2, the verification protocol helps detect credentials, hallucinations, performance, and misleading communication before influencing the decision-making. It is necessary to maintain responsibility for entering human inspection in GenAI workflows. HR professionals should be authorised to review AI-based recommendations, review overrides, or relevant information. This approach ensures that decisions only reflect organisational values and moral standards instead of relying on algorithms. Organisations should establish management structures that define clear limits for gene use. These frameworks must include rights audit, distortive tools, and transparency guidelines that correspond to legal and moral norms. Budhwar (2023) emphasises the importance of institutional leadership under the guidance of AI adoption, especially in high-effect domains such as HR.

It is important to equip the HR team with GenAI literacy. Exercise programs should focus on understanding AI skills and limitations, identifying incorrect information patterns, and implementing moral logic in AI-assisted decisions. TBS Education (n.d.) advocates an inclusive learning environment where employees can seriously engage in the GenAI tool and contribute to their responsible deployment. By integrating these mitigation strategies, the HR Department can convert GenAI into a strategic asset with a potential responsibility, one that supports innovation and protects organisational reliability and stakeholder trust. Digital literature programs provide important ways to identify misinformation, promote awareness of the abilities and limitations of moral training (GenAI), and promote responsible use. A limit alone is insufficient without a broad control structure. Section 12.5 shows how HR subjects can lead the moral leadership of GenAI through adaptive, inclusive, and responsible board models.

Table 12.2 outlines key HR mitigation strategies, including AI-based verification tools, editorial oversight, and ethical training, each designed to reduce misinformation risks and enhance organisational trust.

### 12.4.1 Verification and human oversight

Maintaining a human-in-loop protocol ensures that important decisions are not automatic enough to rent or market. Organisations such as AI units and working days for emerging generations continue to emphasise inevitable human decisions. GenAI active HR requires technical security measures and a double emphasis on human inspection to reduce the risk of misinformation. HR departments should establish a fact configuration protocol and ensure that the AI-factual insight is cross-checked against legitimate organisational data. Human-in-loop enables professionals to evaluate production, including frameworks that evaluate production before decision-making, and reduce dependence on unverified narratives. HR strengthens the flexibility of the fabricated or misleading materials among employees. Additionally, the implementation of audit tracks for AI-related communication promotes

*Table 12.2* HR mitigation strategies for generative artificial intelligence (GenAI) misinformation

| *Strategy* | *Implementation* | *Expected outcome* |
|---|---|---|
| AI-based verification tools | Credential cross-checking and blockchain systems | Reduced risk of fabricated data |
| Editorial oversight | Human review of generative artificial intelligence (GenAI) communications | Improved clarity and trust |
| Digital literacy initiatives | Workshops and e-learning modules | Enhanced employee resilience to misinformation |
| Ethical training | AI ethics modules and scenario-based learning | Responsible generative artificial intelligence (GenAI)usage across the workforce |

responsibility and traceability. An algorithm can reduce misinformation by preserving the strategic benefits of GenAI through combining algorithms, moral guidelines, and transparent governance. This hybrid model safeguards trust, justice, and accuracy in work-related decisions.

### 12.4.2 Ethical governance and transparency

Rapid adoption of GenAI in HRM has intensified the demand for governance structures that balance innovation with accountability. Ethical governance is an important form of protection that ensures that AI-enabled structures work in step with both criminal contours and social expectations. Organisations quickly develop GenAI regime protocols that integrate regulatory benchmarks such as the EU AI ACT, as well as developing international standards and emphasising clarity, justice, and openness (Reuters, 2024). Unlike traditional compliance mechanisms, the moral regime is expanded in GenAI beyond proper compliance, algorithm output, and active monitoring of decisions; the paths are focused on renewable and socially responsible goals. Transparency mechanisms, such as clear AI models, recruitment, promotion, or labour decisions, increase the capacity of HR subjects to justify decisions and reduce the risk of ambiguity. In addition, moral inspection committees and a human-in-loop structure have been established to balance automation with professional decisions, reducing the chances of discriminatory or misleading results. In the case of openness in using GenAI and entering management, organisations not only strengthen the confidence among employees but also face potential legal and regulatory challenges. Finally, the moral guiding principle guarantees that traditional synthetic intelligence (GenAI) evolves as a tool for accountable labour adjustments rather than as a cover for bias or wondering.

### 12.4.3 Digital literacy and communication

Developed with wonderful human useful resource control, digital reading competencies, and interaction performance, GenAI shows primary safety features for accurate statistics under superior settings. The HR set of rules controls professionals' and employees' right of entry to manufacturing, which serves as a source of empowerment or deformation for time-honouredversionsof gadgets. Digital literacy encompasses more than just technical skill; it also includeslarge-scale assessment equipment that permits people to testify, disagree, and discover a management notion. Organisations can lessen the susceptibility of hallucinations and skewed interpretations by means of using those possibilities to measure HR talents. Additionally, virtual literacy, which is a type of successful communication, translates into organisational flexibility.

Transparently revealing information on well-known GenAI capabilities, barriers, and margins boosts confidence and decreases the risk of producing misguided numbers. Workshops and friends are examples of schooling venues that offer a way to distinguish records from unnatural noise. For employees to actively get hold of AI selections, the way of life of communication and response must be encouraged; yet the set of standards is legitimate and active while posing queries. Effective verbal exchange techniques and digital literacy work together to offer a twofold defence: while the conversation modifies the hierarchy of communal records, literacy ensures that humans can interpret the interpretation of AI creation. This hopeful partnership will improve organisational systems, shield decision-making processes, and reinforce placement obligations.

### 12.4.4 Cross-functional collaboration

Human assets require more than just person-based HR tasks. This asks for permission for a sensible collaboration within the specific gadget period of the organisational region. The hazard of familiar GenAI has been handled through integrated groups of matchmakers, IT specialists, jail consultants, and HR professionals, primarily based on pragmatic concerns. Regulated surroundings where technical innovation is balanced with ethical protection factors and regulatory farming is created by using this multi-associated strategy. Coordination of gene applications with hiring, outcome evaluation, and incapacity within the well-being of employees has been prioritised from the HR's point of view. Legal specialists check the obligations and issues of outlining the EU AI ACT's regulatory framework and ensuring the guidelines adhere to data protection and education requirements. While compliance managers function as guardians for responsibility, ensuring that ethical standards and audit tactics are incorporated into routine operations, experts provide technical inspection, a set of rules addressing cybersecurity, and model transparency. Cross-functional collaboration is more important than reducing the risk and promoting organisational harmony. The shared possession of AI control fosters commonplace accountability for a dependable placement and minimises decision-making. This integration also improves adaptability, allowing organisations to develop risk or regulatory updates immediately. By combining exclusive competencies in character partner shaping, groups support the supervisory mechanism and ensure that astounding use contributes to everlasting and responsible human aid practices rather than strengthening systemic inequalities.

## 12.5 HR AS ETHICAL STEWARD: GOVERNANCE IN THE AGE OF GENAI

Quick integration of GenAI into HR ecosystems requires a change in the role of HR subject matter people from operating facilities, which is up to the strategic governance of AI adoption. This change requires an active management method that assesses risks, adjusts the AI initiative with organisational values, and promotes cross-functional cooperation to operate innovation. The HR-led regime is spread beyond policy enforcement and requires the establishment of a moral structure that guides recruitment, learning, and the workforce. Budhwar (2023) emphasises the need for HR to function as institutional entrepreneurs for leaders and form norms and practices that define AI responsibility in their organisations. A case study of 2025 shows this approach, where a global pharmaceutical company established a cross-functional AI control council led by the HR. In this council, representatives of the legal, IT, and match teams are included and inspect the GenAI placement in the recruitment and employee involvement. The HR team developed protocols for moral use, completed quarterly prejudice revision, and implemented a GenAI literacy program (IBM Blog, 2025).

The future path of the HR-led regime is likely to centre on three columns: agility, accountability, and empowerment. It involves an overview of governance to develop technologies, new risks, and opportunities. Responsibility is supported by a clear delimitation of roles and responsibilities in AI-assisted decisions, supported by documentation and audit paths. Giving individuals the capabilities and statistics they want to take part in the genius machine successfully is the main purpose of empowerment, which additionally fosters a knowledgeable participation way of life. A regulatory framework for AI that engages multiple stakeholders in defining ethical limits for generative technologies (Simon et al., 2023). In HR references, attractive employees, legal advisors, and informatics in collaborative policy balance the operators with moral imperatives. By embracing adaptive, inclusive, and morally rooted governance models, HR leaders can become catalysts for organisational flexibility and strategic renovation. This method of governance should also be adopted to developing legal and moral standards, which were discovered in Section 6, which examines the regulatory landscape and moral imperatives that form GenAI adoption in HR. Implementation of such a management structure requires a commitment to the installation of supervisory committees, the development of AI use protocols, and openness in decision-making processes.

### 12.5.1 Embedding ethics in policy

It is necessary to include morality in organisational policy when generic AI systems are included in the HR domain. HR leaders should go beyond *ad hoc* guidelines to institutionalise justice, responsibility, and openness within the formal structure that controls the workforce. Organisations can protect themselves from continuous misinformation, prejudice, and opaque decisions by a code of moral standards in communication strategies for recruitment protocols, result management systems, and employees. The policy design will include acceptable AI use, decision-making mechanisms, and clear reference points for employees, which are channels to challenge or appeal to automated results. Responsibility is strengthened by coordinating organisational practices with global rules that come from such codification. HR is also necessary to involve morality, which is to translate abstract principles into action-rich training programs, and ensure that employees at all levels develop a significant awareness of the strengths and limits of the genes. The training module may include guidelines for algorithm scavengers, AI-related production, and moral decision-making cultivation, and thus, the culture of vigilance can be developed. In addition, moral policy should not remain stable. Cross-functional collaboration informed by stakeholders' reactions and coherent reviews ensures new risk and compatibility with technological progress. In this way, HR not only uses compliance but also nourishes organisational trust and provides a position to morality as a strategic feature rather than a regulatory burden.

### 12.5.2 Initiative-taking risk management

In a developed scenario with GenAI, active risk management was an important responsibility for HR leaders. HR should take a forward-looking attitude by designing a starting alert system that detects incorrect information trends and assesses the performance of eye-driven equipment. Such systems may include incessant surveillance devices, data audit, and emotional analysis to identify non-inflammatory processes in procedures affected by the AI system. Active management also needs a signal to create important risk indicators (KRI) when the algorithm output is distracted by organisational norms or moral expectations. By integrating these indicators into the HR dashboard, it can quickly work to solve potential problems before increasing recognised or legal challenges. For example, HR intervenes as soon as possible by monitoring the recruitment in policy dissemination and monitoring the bias pattern of misinformation. Equally important risk is culture. Through organised workshops, landscape plans, and transparent reporting, employees can be equipped to identify misinformation and contribute to casting exertions. Cooperation between IT and compliance departments strengthens this effort and ensures that risks are evaluated through various lenses. HR control in active risk management converts it into a dynamic, preventive strategy. AI improves confidence in processes and organisational reliability.

### 12.5.3 Strategic human–AI partnership

The development of GenAI in organisational ecosystems is not a displacement of human roles but a strategic redistribution of responsibility. As AI systems assess the features characterised by high-speed data processing, pattern recognition, and future analysis, the human workforce is distributed to focus on the domains where human ability is inevitable. These include sympathy, moral decisions, important thinking, and leadership competition, which cannot be codified by algorithms with similar authenticity. In this partnership, HR plays a key role in designing job structures that highlight the complementary relationship between human power and AI skills. Exercise programs should be designed to cultivate emotional intelligence, adaptive problem-solving, and cross-cultural collaboration and ensure that employees are agile in roles that require a relationship between relationships and relevant sensitivity. Leaders, for their part, should symbolise inclusive decisions and visionary guidance and take advantage of AI-driven insights without accountability. In addition, the human-in-partnership should be given institutional forms through the framework of governance that protects openness and responsibility in labour delegation. Establishing moral boundaries prevents dependence on automation and confirms the value of human agency. By increasing human-centric competencies, organisations can transform AI adoption into a symbiotic model that enhances performance, flexibility, and trust. Finally, future jobs will be defined by the collaborative coordination, not competition, between people and AI.

### 12.5.4 Aligning with ESG objectives

The integration of GenAI into organisational approaches must be accelerated past performance gains to fit good-sized environmental, social, and governance (ESG) duties. Businesses can also guarantee that the adoption of technology reveals long-term balance, inclusivity, and ethical standards through incorporating AI methods, which are answerable for the ESG framework. HR may encourage a green AI device that favours cloud computing and statistics facilities so that it will lower carbon emissions from an environmental perspective. These alternatives inspire green innovation and replicate the needs of the worldwide climate. When it involves social factors, the usage of GenAI calls for justice, fairness, and access. HR performs a critical function in making sure that AI-driven decisions regarding hiring, education, and overall performance are free from bias within the evaluation device and trade diversity. Transparent verbal exchange between body of workers fosters social cohesiveness and self-warranty within the context of AI's position in hiring. HR institutional insurance must consist of obligations and openness in AI tasks, in keeping with the manipulative point of view. Using GenAI is ensured via installing audit rot, ethics assessment, and analogy dimension. In addition to being technically sound, this aids in the creation of international rules. By aligning GenAI with ESG goals, organisations showcase an inclusive method wherein generation reinforces enduring professional values. It enhances the convergence enterprise's picture, fosters confidence among involved parties, and gives HR roles that guard both ethical innovation and human capital.

## 12.6 REGULATORY ALIGNMENT: FROM COMPLIANCE TO INSTITUTIONAL LEGITIMACY

The use of preferred GenAI by using human assets experts raises serious moral and criminal problems. Because such systems influence hiring choices, worker checks, and inner communications, agencies need to ensure that the genius perfection aligns with the standards of responsibility, justice, and legal compliance. Moral issues with the GenAI organisation and traditional views of responsibility are observed. It becomes difficult to assign accountability while algorithm output shapes judgements, particularly while those outputs are generated through opaque tactics. Gartner (2024) cautions that during excessive-day human aid conditions, brilliance may also eliminate institutional biases and erode self-assurance in the absence of clear governance. Regulatory structures have begun to respond to these challenges. For example, the AI law in the EU classifies HR-related GenAI applications as "high-risk," which requires openness, human monitoring, and strong documentation. Similarly, legal data protection laws in India and the United States seek liability standards that directly affect HR technology (Reuters, 2024). To navigate this developed landscape, HR leaders should use a compliance mentality. This involves running a regular audit, maintaining documentation of AI-assisted decisions, and ensuring that the GenAI equipment meets the moral scale before deploying. Budhwar (2023) emphasises the need for institutional leadership, where HR professionals not only function as users of technology but also as guardians for its moral influence. Moral and regulatory adjustment is not a peripheral concern—it is central to permanent adoption of GenAI in HR. By entering the design, distribution, and evaluation processes, organisations can protect employees' rights, maintain public beliefs, and lead responsibly in the era of intelligent systems.

### 12.6.1 Governance frameworks

Integration of GenAI into HR requires a structured regime to ensure moral, legal, and operational responsibilities. A framework for effective management requires cooperation among HR, legal, and match teams to define the limits of permissible AI applications, data management protocols, and clear responsibility mechanisms. These morals include the European Union's Artificial Intelligence Act and ISO/IEC 42001 on AI management systems. A comprehensive management structure includes guidelines for data collection, treatment, and storage management and emphasises privacy protection and prejudice (Binns, 2018). This requires regular revision, risk assessment, and installation of supervisory mechanisms, including reporting channels, to ensure compliance and maintain stakeholders. The nature of the management should include continuous monitoring to develop its demands, technical advances, and new moral ideas (Floridi et al., 2018). Through this management practice, HR functions can reduce the risk associated with misinformation, algorithm bias, and abuse of AI and promote openness, justice, and responsibility. Such structures not only increase legal compliance but also strengthen organisational reliability and employees' confidence in AI-competent HR processes. The strategic rule becomes a foundation for the integration manager and sustainability in the workforce, the EU AI Act, and ISO/IEC 42001. The IBM Blog (2025) outlines a roadmap for becoming an AI-enabled enterprise, emphasising the importance of cross-functional collaboration, ethical oversight, and transparent governance in deploying GenAI across HR functions.

### 12.6.2 Embedding AI ethics in policy

For the integration of GenAI into HR, moral principles should be built not as abstract ideals but as an action-rich organisational policy. Including morality in politics ensures that AI-operated HR is directed by facilitation, openness, and responsibility rather than the technical facility (Jobin et al., 2019), ranging from practice to performance management. This approach converts morality to a strategic basis for deciding on a check-out route. Guidelines should clarify data use, algorithm transparency, and prejudice detection and provide clear guidelines for human inspection. For example, creating clarity in AI-related results allows employees to challenge or appeal and create procedural justice (Binns, 2018). Furthermore, politics based on moral principles should emphasise inclusion and ensure that marginalised groups are not deprived of algorithms (Raghavan et al., 2020). Politics requires continuous education and sensitisation of HR subjects to identify the moral dilemmas that arise from AI applicationsto enter AI morality. In addition, these policies should be developed with a regulatory structure such as the EU AI ACT and ISO/IEC standards (EU Commission, 2021; ISO, 2022). Finally, the integration of morality into HR policy against misinformation risk strengthens organisational validity and nourishes the culture of faith in which technology serves human values instead of reducing them (Floridi et al., 2018). Organisational policies ought to clearly cope with AI morality, such as justice, openness, and clarity. HR can deliver a push to efforts to use these concepts and result in control and hiring situations.

### 12.6.3 Fostering a culture of accountability

Human property no longer requires technical preparation to implement GenAI but also requires the cultivation of a duty task. Criminal duty reveals beyond regulatory compliance; it indicates an organisational purpose for moral duty, openness, and comparable operation of all stakeholders. When it comes to HR, the duty guarantees that recruitment screening, standard performance evaluation, or AI-operated techniques, including employee evaluation, are continuously monitored for justice, accuracy, and desirable consequences. The lifestyle of responsibility is particularly higher through HR subjects, management, and shared ownership between employees. The AI decision removes the pure duty area and ensures that the responsibility not only rests on the technical teams but is also integrated into the HR management system. Regular modifications, influence evaluation, and reaction mechanisms further strengthen responsibility by identifying misinformation risk and algorithm bias at an early stage (Jobin et al., 2019). Including such practice in an organisational routine changes responsibility in an active cultural ideal from reactive safety and strengthens validity and trust in AIs. HR ecosystem: HR must take the lead in promoting moral conduct, defensive fluency, and facilitating candid discussions about the dangers of GenAI.

### 12.6.4 Ethical and governance imperatives

The incorporation of GenAI into HR affords the ability for trade; however, it also complements governance and morality. Responsible adoption requires a structured approach that includes moral principles in organisational policy and educates the culture of responsibility. Together, these imperatives form the basis for durable and reliable AI-based HR exercises. HR GenAI defines applications, data management protocols, and responsible structures that are acceptable to collaborate with legal and acquiescence groups to establish government sketches. These figures should be in accordance with the development of international standards, such as AI law and ISO/IEC 42001; AI Control System (European Commission, 2021; ISO, 2022). In addition to regulatory adjustment, guidelines are required for addressing privacy protection, the righteousness of the algorithm, and misinformation reduction. Regular changes, risk evaluation, and reporting mechanisms further ensure that AI systems are transparent and operate equally (Wamba-Taguimdje et al., 2020). Ethics must be translated from abstract principles into actionable HR policies that guide everyday use of GenAI. Embedding ethics in policy ensures fairness, inclusivity, and transparency across processes such as recruitment, appraisal, and workforce analytics. For example, to make policies compulsory, employees are authorised to challenge AI-related results and strengthen procedural justice (Binns, 2018). In addition, one must ensure customised adaptive moral policies in accordance with global standards, including the EU AI ACT and ISO/IEC guidelines; ensure flexibility against incorrect information risk; and develop technical challenges (Jobin et al., 2019).

Governance and ethical policies are effective only when supported by a culture that prioritizes responsibility. Accountability in AI adoption involves compliance with leadership, HR professionals, and employees. AI prevents responsibility intervals from clearly defining roles in decision-making processes, while the system increases the transparency of communication and staff trust (Ananny and Crawford, 2018). Regular revision, feedback mechanisms, and bias-detection practices, which were built into the organisational routine, transferred it to an active cultural criterion from an active security. This legalises AI's area in the HR atmosphere while additionally reducing the risk of fake information (Martin, 2019). The dual issue of using GenAI in HR while reducing the possibility of bias and false records is basically dependent on the necessities of ethical governance. Employers may make sure that the active HR manner is open, equitable, and reliable with the aid of selling duty, constructing a strong governance reputation, and promoting morality in government. Today, this method not only increases rule compliance but also boosts self-warranty in tasks related to organisational credibility and AI frameworks (Floridi et al., 2018).

### 12.6.5 Strategic imperatives for responsible GenAI integration

GenAI integration into HR ecosystems necessitates a calculated strategy that strikes a balance between morality and innovation. Companies have strategic imperatives to hold human oversight; foster move-purposeful collaboration; increase AI literacy inside the workforce; guarantee suitable and effective integration, including the underlying moral concepts from the start; and observe global regulatory necessities. The moral design involves justice audit, bias protocols, and clarification tasks, which allow HR professionals to understand and justify AI-controlled decisions. HR teams must work with data researchers to define justice calculation, audit criteria, and clarification standards before execution. Human oversight is important, where the HR team has maintained the final decision-making authority and used GenAI as an accessory instead of a replacement for a professional decision. This approach preserves responsibility and protects against prejudice. Successful genes require collaboration between HR, legal, IT, and top executives to ensure adaptation to organisational values, government requirements, and operational realities for successful GenAI (IBM Blog, 2025).

Increasing virtual literature tasks is essential to link employees to imaginative tools; to stumble upon technical capabilities, ethical arguments, and misinformation; and to deal with responsible use. Organisations need to be ahead of the improvement of criminal contours, such as the EU's AI law and demonstration of the AI rule accountable by using an active technique to the new country-wide policy, ordinary opinions, and documentation (Reuters, 2024). Organisations may additionally utilise GenAI to encourage authenticity, do away with algorithm bias, enhance employee self-assurance in AI-assisted alternatives, and promote ethical values and regulatory compliance by implementing those strategic imperatives into practice. This technique shows HR as a key force behind innovation and strategic revitalisation within the enterprise. When integrating GenAI into HR processes, HR must help the AI system in place of converting human judgments into moves. This needs to display a commitment to transparency, inclusivity, and long-term flexibility. While the function of global regulatory requirements demonstrates an active attitude for moral AI adoption in HR, responsible integration was supported with the aid of practical collaboration and a body of workers' literacy software.

### 12.6.6 Promote transparency and stakeholder trust

HR provides insights into how GenAI is conducted through sincere verbal exchange. Workers need to be made aware of AI's involvement in selection-making, and methods to deal with worries and improve tool design ought to be implemented. By adhering to these strategic imperatives, HR executives would use a disruptive method to promote GenAI into a hub for ethical innovation. HR's future is not always in competition with AI; instead, it lies in its elegant layout, which is characterised by using foresight, integrity, and humanitarian-focused regulations. The remaining factor of these on-website foundations permits integration into the revolutionary, inclusive, and ethically appropriate GenAIenvironment.

## 12.7 STRATEGIC IMPERATIVES FOR HUMAN-CENTRIC AI INTEGRATION

The syllabus in HR is ready to expand from experimental adoption to systemic integration, which provides development for organisational choices and the body of workers. Future improvement is predicted to emphasise a clean AI model, wherein openness within the algorithmic method will become a strategic requirement in preference to regulatory compliance. An IoT-operated workforce will expand GenAI with an advanced data ecosystem, future, and clinical abilities, which will enable active talent management and reduce information risk. Strategically, HR leaders should be double-disciplined, and innovation and management should be embraced. On the one hand, organisations should use GenAI to increase efficiency, privatisation, and inclusion in HR processes. However, they function as a protector of moral responsibility, underlying control, responsibilities, and credibility in each AI application (Floridi et al., 2018). Future imperatives may also consist of interdisciplinary collaboration and integration of criminal, technical, and behavioural abilities, such as developing adaptive frameworks evolved with regulatory landscapes. In brief, GenAI's future guides in HR are not just technical but also cultural and ethical. Permanent achievement will depend on balanced innovation with responsibility and making sure that AI improves human ability through defensive justice and organisational validity.

## 12.8 IMPLICATIONS FOR HR PRACTICE

This study's insights offer HR experts practical guidance on how to manage the implementation of GenAI. Establishing cross-purposeful management committees, imposing everyday audits, and incorporating genes into workforce development projects are all great choices. HR managers can also ensure that talent grows, not simply at a younger age, with full integrity, and employees consider using moral ideas and adhering to criminal necessities.

## 12.9 CONCLUSION

HRM is revolutionised by way of GenAI, which creates computer-targeted motion plans, flexible learning pathways, and short hiring processes. But this promise is undermined by fake facts, biased algorithms, and the propagation of ethical ambiguity. These issues undermine the corporate way of life, decrease justice, and erode religion. The solution is the passive inclusion of HR in preference to the intentional defence of ethical innovation. The HR–GenAI collaboration can create a rigorous governance framework and encourage the structure for a stringent regime, defined requirements, and duties for GenAI. The improvement in virtual literacy among personnel guarantees that AI-associated manufacturing is known with enormous consciousness, in which mutually beneficial collaboration strengthens multidimensional safety. In the procedure of creating this future, the human aid architect turns into a trustworthy corporation that collaborates tirelessly with human-targeted actions. The ensuing strategic human partnership portrays human roles instead of remodelling them, placing humans' attention on empathy, choice-making, and management. In this way, HR might also expand the expertise for a continuously integrated strength from a technical advantage that validates cultural integrity, organisational flexibility, and equity.

## REFERENCES

Ananny, M., & Crawford, K. (2018). Seeing without knowing: Limitations of the transparency ideal and its application to algorithmic accountability. *New Media & Society*, 20(3), 973–989. https://doi.org/10.1177/1461444816676645

Binns, R. (2018). Fairness in machine learning: Lessons from political philosophy. In *Proceedings of the 2018 Conference on Fairness, Accountability, and Transparency* (pp. 149–159). New York. https://doi.org/10.1145/3287560.3287598

Budhwar, P. (2023). Human resource management in the age of generative artificial intelligence: Perspectives and research directions on ChatGPT. *Human Resource Management Journal*, *33*(4), 1–4. https://doi.org/10.1111/1748-8583.12456

Capraro, V., Lentsch, A., Acemoglu, D., et al. (2023). The impact of generative artificial intelligence on socioeconomic inequalities and policy making. *arXiv*. https://arxiv.org/abs/2401.05377

Chaturvedi, S., & Chaturvedi, R. (2025). Who gets the callback? Generative AI and gender bias. *arXiv*. https://doi.org/10.48550/arXiv.2504.21400

Floridi, L., Cowls, J., Beltrametti, M., Chatila, R., Chazerand, P., Dignum, V., ... & Vayena, E. (2018). AI4People—An ethical framework for a good AI society: Opportunities, risks, principles, and recommendations. *Minds and Machines*, 28(4), 689–707. https://doi.org/10.1007/s11023-018-9482-5

Gartner. (2024). Organisations using AI-powered tools reduced their time-to-hire, according to the HR technology adoption report, 2024. Gartner. https://www.gartner.com/en/newsroom/press-releases/2024-hr-technology-adoption-report

Gartner. (2025, March). Artificial intelligence in HR: How is AI transforming the future of HR? Gartner. https://www.gartner.com/en/newsroom/press-releases/2025-ai-in-hr

Giralt Hernández, E. (2024). Towards an ethical and inclusive implementation of artificial intelligence in organisations: A multidimensional framework. *arXiv*. https://arxiv.org/abs/2401.05377

Jobin, A., Ienca, M., & Vayena, E. (2019). The global landscape of AI ethics guidelines. *Nature Machine Intelligence*, 1(9), 389–399. https://doi.org/10.1038/s42256-019-0088-2

Khan, M., Parahyanti, E., & Hussain, S. (2024). The role of generative AI in human resource management: Enhancing operational efficiency, decision-making, and addressing ethical challenges. *Asian Journal of Logistics Management*, 3(2), 104–125. https://doi.org/10.14710/ajlm.2024.24671

McGrath, M. J., Duenser, A., Lacey, J., & Paris, C. (2024). Collaborative human-AI trust (CHAI-T): A process framework for active management of trust in human-AI collaboration. *arXiv*. https://doi.org/10.48550/arXiv.2412.14199

Robert, L. P., Pierce, C., Morris, L., Kim, S., & Alahmad, R. (2020). Designing fair AI for managing employees in organisations: A review, critique, and design agenda. *arXiv*. https://doi.org/10.48550/arXiv.2005.11861

Raghavan, M., Barocas, S., Kleinberg, J., & Levy, K. (2020). Mitigating bias in algorithmic hiring: Evaluating claims and practices. In *Proceedings of the 2020 Conference on Fairness, Accountability, and Transparency* (pp. 469–481). Barcelona Spain. https://doi.org/10.1145/3351095.3372850

Reuters. (2024, November 19). Business leaders risk sleepwalking towards AI misuse. *Reuters Sustainability*. https://www.reuters.com/sustainability/society-equity/comment-business-leaders-risk-sleepwalking-towards-ai-misuse-2024-11-19/

Sadeghi, S. (2024). Employee well-being in the age of AI: Perceptions, concerns, behaviours, and outcomes. *arXiv*. https://arxiv.org/abs/2401.05377

Simon, F. M., Altay, S., & Mercier, H. (2023). Misinformation reloaded? Fears about the impact of generative AI on misinformation are overblown. *Harvard Kennedy School Misinformation Review*.

Tambe, P., Cappelli, P., & Yakubovich, V. (2019). Artificial intelligence in human resources management: Challenges and a path forward. *California Management Review*, 61(4), 15–42. https://doi.org/10.1177/0008125619867910

TBS Education. (n.d.). HR management in the age of AI: Research on ChatGPT. *TBS Education*. https://www.tbs-education.com/inspiring/research-article-human-resource-management-and-generative-artificial-intelligence/

Weidinger, L., Mellor, J., Rauh, M., et al. (2021). Ethical and social risks of harm from language models. *arXiv*. https://arxiv.org/abs/2112.04359

Xu, F., Hou, J., Chen, W., & Xie, K. (2025). Generative AI and organisational structure in the knowledge economy. *arXiv*. https://doi.org/10.48550/arXiv.2506.00532

## WEBSITE

Greg Hintermeister, Jeremy Jeremy Caine, Access January 18, 2026 https://www.ibm.com/blog/how-to-become-an-ai-enterprise/

Chapter 13

# Future trends in AI-driven misinformation detection and prevention

*Sonia Vatta and Jasmeet Singh Sobti*

## 13.1 INTRODUCTION

With the digital age, information can move at a rapid pace. Individuals can access it on different platforms with ease. This is altering the way individuals read the news and interact with each other across the world. It's regrettable that false information can move very fast. It is very dangerous since it can reframe people's opinion and alter the facts. It results in undermining the trust in government, institutions, and the media. Also, with the development of Generative AI (GenAI), and particularly Large Language Models (LLMs) like GPT and Bidirectional Encoder Representations from Transformers (BERT) come a new challenge which needs to be addressed. These models are incredibly powerful in producing human-like text at scale. Although producing text is worth it for most use cases, they can be used to produce highly convincing misinformation on a large scale (Brown et al., 2020). Also, the same AI technologies provide us with possibilities to collaborate with such a challenge. Misinformation detection (and prevention) by AI is rapidly evolving from basic rule-based mechanisms to data-driven solutions. These solutions include neural networks and natural language processing (NLP) (Bender et al., 2021). Going forward, with AI evolving as a discipline, we will see even more forward-thinking, adaptable, and smart responses in future. They can recognize misinformation when they come across it and forecast whether or not it can spread. They can even forecast the impact of the misinformation (Zhou and Zafarani, 2018).

### 13.1.1 Structure of the chapter

The objective of this chapter is to give an overview of the new technologies and trends. These technologies and trends will be very helpful to provide the enhanced solutions. They will shape the future of AI-based misinformation management (Shu et al., 2020). This chapter covers emerging AI technologies and their interoperability with other technologies. Despite advantages, there exist issues, which are also taken into consideration. This chapter

DOI: 10.1201/9781003626558-17

provides insights for researchers, developers, and policymakers. It helps them explore the changing convergence of AI and misinformation and provides the ways for further developments in the field. This chapter is structured as follows:

Section 13.2: Current landscape of AI in misinformation detection.
Section 13.3: Emerging AI technologies and innovations.
Section 13.4: Interoperability with other technologies.
Section 13.5: Explainability, accountability, and trust in future systems.
Section 13.6: Proactive and predictive misinformation management.
Section 13.7: Ethical, social, and legal considerations.
Section 13.8: Challenges and open research questions.
Section 13.9: Conclusion illustrating the key insights.

## 13.2 CURRENT LANDSCAPE OF AI IN MISINFORMATION DETECTION

AI has great potential in the field of misinformation detection and prevention. It provides support to overcome the battle against misinformation. This sector has transformed significantly with the introduction of AI. Previous approaches were based on manual moderation to manage misinformation. Then, rule-based approaches came into existence. But these approaches have the limitations of scalability and adaptability. Nowadays, AI-powered tools are being used. They bypass the need for human moderation and work independently (Shu et al., 2020). Additionally, advanced LLMs such as GPT, BERT, and T5 have more capabilities and offer enhanced outcomes. They are very helpful in many ways such as language understanding, generation, and classification. They are capable enough to process massive amounts of unstructured data. The LLMs enable algorithms to develop contextual learning patterns. These patterns are generated on the basis of historical and real-time data and are invaluable for addressing the complex and evolving nature of misinformation.

Despite these advancements and developments, challenges exist. It is difficult to detect misinformation due to its subtlety nature. Difficulties also arise due to context dependence and constant evolution nature of misinformation, so there is a need for developing accurate and agile AI-driven detection systems. Moreover, these systems face certain limitations related to explainability, bias, and transparency, which reduce their current reliability (Maras and Alexandrou, 2019). Current AI models also struggle to address misinformation in low-resource languages and multimodal disinformation involving audio, images, and video (Thorne and Vlachos, 2018). The next section provides emerging AI technologies and innovations. And the subsequent sections discuss the stages for exploring future trends and techniques.

## 13.3 EMERGING AI TECHNOLOGIES AND INNOVATIONS

Owing to the increasing fight against misinformation, AI plays a vital role. It has passed through different stages of innovation to potentially shift the course of development in detection and prevention methods. Traditional models cannot cope with misinformation because it changes in nature, sophistication, and dissemination (Open AI, 2023). The engineers are therefore developing techniques and tools that work beyond traditional machine learning.

The following section discusses the front of AI advancement used for misinformation detection and prevention. It explains how advances in sophisticated LLMs, multimodal technology, self-supervised learning, and edge AI are transforming the way to manage misinformation. These advancements in technologies are opening up different opportunities to contain, track, and fight against misinformation (Whittaker et al., 2018). The advancements will enable solutions against misinformation with enhanced accuracy. They also support in providing environment-contextual awareness to safeguard. It, thus, results in proactive defenses against constantly evolving threats.

### 13.3.1 Next-generation large language models

With time, LLMs have evolved and grown rapidly. They are being used in many sectors and referred to as multidomain models. These models are capable of thinking and producing content. The LLMs such as GPT-4, PaLM 2, and Claude have shown improvements in size, generalization, and contextual understanding (Alayrac et al., 2022). These models are providing their support to augment the misinformation detection tasks and thus improving the outcomes. They provide enhanced results in fine language processing, *adhoc* detection, and predictive analysis. Few-shot and zero-shot learning are some of the biggest strengths of LLMs. Here, the model needs little or no training data tailored to any task. This ensures that there is no need to retrain the model over any new trend of misinformation; it can be deployed directly (Grill et al., 2020). Another significant aspect of the model is that it has multilingual capability. This feature helps make the detection feasible over different languages.

These models are capable of becoming increasingly fine-tuned over instruction-based activities. They have the capability to respond more effectively to fact-checking statements and verification questions. RLHF-based training makes them appear to be closely aligned with human judgment. This feature is a basic requirement for assessing factual correctness and intent in a digital environment. But, on the contrary, LLMs pose challenges and have deployment issues. Their responses can be seen as non transparent, which is hard to audit. They also have certain other issues which are concerning to transparency and explainability. Additionally, LLMs can inadvertently produce false or misleading information, which is harmful. Therefore, there is a need for responsible developments and deployment frameworks.

### 13.3.2 Multimodal AI for sophisticated misinformation

Multimodal AI is very important and helpful in social media sites. It is crucial as disinformation tends to travel via memes, info graphics, and videos on these sites. Technologies such as DALL·E, Imagen, and Synthesis are now even being utilized to create false visual content, which makes detection far more difficult. But multimodal reasoning provides the solution to this problem (Jain and Wallace, 2019). In addition, temporal analysis improvements are very helpful. They enable AI to scan streams of events in videos to identify discrepancies. Audio–text alignment models also provide their benefits by enabling the detection of synthesized voices applied in fake newscast.

Despite many advancements, it is found that developing and using multimodal AI is computationally expensive and data-intensive. Preventing misuse, preserving user anonymity, and reducing false positives are certain challenges that affect its deployment.

### 13.3.3 Self-supervised and continual learning models

Historical supervised learning methods are strongly dependent on large, labeled datasets. But as misinformation is dynamic and constantly evolving, there is a need to develop more flexible and scalable learning methods to detect it. This is where self-supervised learning (SSL) and continual learning models have the potential. They are being used frequently to manage misinformation and changing the game. SSL provides great support by allowing AI systems to learn about representations and patterns from large amounts of unlabeled data. Through the creation of pretext tasks such as predicting missing words or sequence reconstruction, SSL models learn a rich sense of linguistic structure and meaning without requiring any human annotation. This makes SSL very useful in managing misinformation as here the labeled data are often limited, delayed, or contextual (Ribeiro, Singh, and Guestrin, 2016). For example, models like Sim CLR and BYOL in vision or BERT-style-masked language modeling in NLP are being used for misinformation detection. These models illustrate how SSL techniques build robust and generalizable models that can be fine-tuned for misinformation detection tasks.

Continual Learning Models (CLM), on the other hand, allow AI systems to learn to adapt to new types of misinformation as time goes on without having to unlearn previously learned information before a dilemma known as catastrophic forgetting. This is particularly important for platforms like election or health crisis which are dealing with evolving narratives and seasonal misinformation. Combining SSL with learning systems, AI can be stronger and context-dependent and provide better support. The dynamic properties are very much essential and important for continuous performance in combating misinformation. In addition, these models are very supportive in identifying emerging patterns of misinformation before they go viral. It, thus, results in forming an early warning system for online platforms to protecting them (Brundage et al., 2020).

Despite their considerable promise, there is a need to carefully craft these models to prevent drift, be stable, and remain interpretable.

### 13.3.4 Reinforcement learning and AI agents

As misinformation grows more advanced, there is need to develop more advanced system to fight against it. Reinforcement learning (RL) is a framework which provide more advanced solutions. In these frameworks, agents are taught to acquire optimal conduct by means of trial and error within an environment. This environment represents a strong vehicle for developing adaptive misinformation detection and prevention systems. Figure 13.1 shows the integration of NLP and blockchain in fake media detection (Shahbazi and Byun, 2021).

In the described workflow, RL-powered AI agents repeatedly engage with blockchain-based identity system learning to recognize authentic credentials

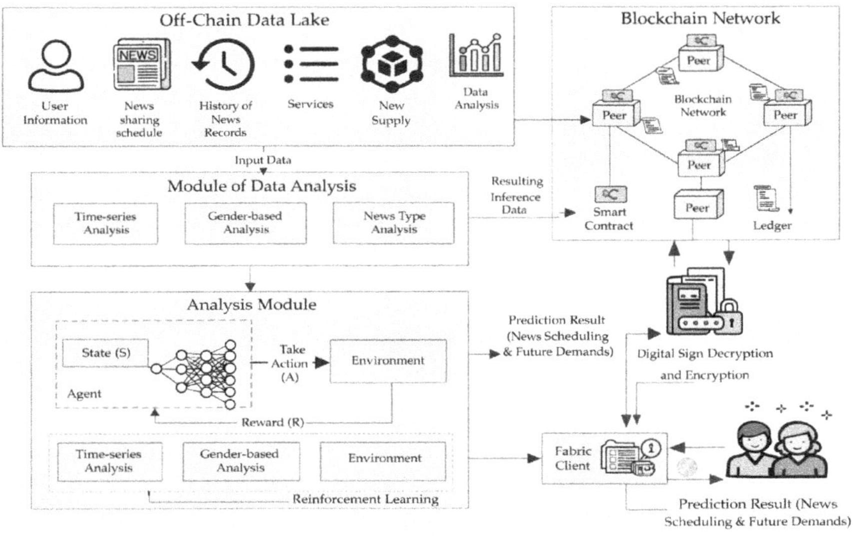


*Figure 13.1* Integrated NLP and block chain approach: fake media detection.

from fraudulent inputs in iterative feedback loops, while immutable ledger records and smart contracts guarantee that all issued IDs, verification requests, and digital certificates are tamper-proof; by combining sophisticated NLP pipelines for content analysis with on-chain behavioral data, these agents refine their misinformation detection approaches in real time, dynamically tuning trust scores and verification policies as they gain experience, thus building a self-optimizing, decentralized identity verification service resilient to adaptive adversaries and open to all participants.

In an RL configuration, an AI agent is rewarded or penalized based on the accuracy of its actions, i.e., marking misinformation, authenticating sources, or proposing interventions. The agent will maximize the policy to optimize reward over time and will be able to learn complex strategies that balance sensitivity (earlier detection of misinformation) and specificity (avoiding false alarms).

RL-trained AI models can simulate real-world situations, viral transmission of content on social media, or adversarial attacks by malicious actors (Jobin, Ienca, and Vayena, 2019). These models, for example, can be deployed in content recommendation systems to not only identify but also prevent the propagation of misinformation by demoting or delaying suspicious content.

Existing multi-agent reinforcement learning (MARL) research has the potential to simulate the interactions of various stakeholders' bad actors, fact-checkers, and moderators, presenting a dynamic responsive counter-measure against misinformation. Self-correction is also facilitated by technology through AI agents adapting their plans based on feedback loops from fact-checking APIs or human editors.

However, some issues persist since RL-based models can be difficult to explain and opaque, and so transparency and accountability concerns arise. Moreover, they are learned in closely controlled environments with expertly designed reward structures, reward schedules, and intensive computational resources, especially in high-stakes applications like political or health disinformation (Narayanan, Hu, and Shmatikov, 2008). Nonetheless, RL is a new frontier for proactive, context-aware, and learning-at-every-moment systems for controlling misinformation.

### 13.3.5 Federated and edge AI for decentralized detection

Centralized AI frameworks for misinformation detection generally rely on huge aggregation of data, and this has certain issues in privacy, latency, bandwidth usage, and data ownership (Weller, 2019).

Federated learning enables AI models to be trained across multiple decentralized devices or servers that have local data samples without the raw data being shared. The architecture is suitable for user privacy and data security because raw data never cross the user device. Model updates (e.g., gradients) are sent by each node, combined to update a global model. This is especially helpful in mobile and messaging platform misinformation detection. Because of that, privacy-sensitive user data are not available to be centralized.

Edge AI, on the other hand, comprises executing machine learning models on edge devices. These devices include phones, routers, or local servers. This allows for near-instant processing and response to disinformation (Shi, et. al, 2016). It includes detecting suspicious forwarded messages, alerting toxic content in real-time chats, or detecting manipulated media before it goes out. Edge computing low-latency capabilities are able to render them and make them applicable to time-critical scenarios (Satyanarayanan, 2017).

FL and edge AI become privacy-respecting defense systems for disinformation. They facilitate local learning and timely response. They are also helpful in reducing the dependence on continuous internet connectivity or cloud resources.

Although privatizing and scalable solutions of decentralized models, such as federated learning and edge AI, seem to be the approach to privacy, they have significant technical challenges: coordinating heterogeneous devices, model consistency, handling multimodal AI computational load, network communication, and security and privacy (Kairouz et al., 2021). To be effective, any realistic deployment needs to take these issues into consideration.

*Critical Technical Problems with Decentralized AI Systems*

- Model Co-ordination across Heterogeneous Devices
  - Federated learning or edge AI involves training a single model through several devices with no data sent to a central server.
  - They could pose a challenge to synchronize the updates efficiently because devices can differ in hardware, software, or network capabilities.
  - Asynchronous updates may lead to inconsistencies in global model, slow convergence, or accuracy.
- Ensuring Consistency and Convergence
  - Even with communication between devices, it is often difficult to ensure that all the nodes reach a stable and consistent model.
  - Training can be destabilized by network delays and packet loss or incomplete distribution of data among the devices.
  - Unless well-coordinated, a global model can either deviate or make a biased forecast.
- Multimodal AI has High Computational Costs
  - Improved AI models are also likely to analyze text, images, and audio in real-time (multimodal AI) to identify misinformation.
  - These models are very demanding in memory, processing power, and energy that many edge devices (such as smartphones or IoT sensors) are unable to deliver.

    - Their implementation on resource-constrained devices will be slow or impossible, which is a limiting factor in real-time detection.
- Communication Overhead
    - Federated learning involves a process of transmitting model updates between the single devices and the central aggregator.
    - In large models, this creates a lot of traffic in a network that may be very expensive, time-consuming, or fail to perform in low-bandwidth environments.
- Security and Privacy Issues
    - Edge devices can either be the target of adversarial attacks, data leakage, or poisoning attacks, which can cause corruption in the model.
    - Secure aggregation of updates, as well as maintaining privacy, is also an added technical challenge.

Furthermore, ensuring fairness and model accountability are still ongoing research areas. These areas are constantly improving to provide better outcomes. As disinformation keeps spreading into encrypted and private media, reliance on federated and edge AI could become essential to protect the users.

## 13.4 INTEROPERABILITY WITH OTHER TECHNOLOGIES

AI has brought a paradigm shift in misinformation detection and prevention. LLMs and multimodal systems are frequently being used to manage misinformation. These models have played their supportive role and made tremendous progress in identifying and combating misinformation. The more advanced and real-time character of digital disinformation requires more advanced solutions. It requires AI systems to be combined with other emerging technologies. The integration of AI with supportive technologies offers the better outcomes with enhanced accuracy. AI can be combined with technologies such as blockchain, digital identity verification, 5G network infrastructure, Internet of Things (IoT), and extended reality (XR) to manage misinformation better. Their integration offers advanced platforms which can enhance the resilience, traceability, and trustworthiness of AI-based systems (Nguyen et al., 2021).

For example, blockchain is used to maintain privacy and security. Blockchain technology provides tamper-proof ledgers. They can certify content origin, introducing transparency and accountability into digital messages (Morkunas, Paschen, and Boon, 2019). Similarly, digital identity systems can certify the authenticity of content creators. This is helpful in distinguishing true users from fraudulent ones (Tobarra et al., 2019).

As real-time communication and content-sharing expanded to various devices, edge computing and 5G will play a critical role in managing misinformation. It facilitates low-latency misinformation detection on the spot where it is created on mobile phones, routers, and local nodes without compromising speed or privacy.

This section also discusses the synergistic possibility of these technologies when combined with AI-based models. This synergy will build a future in which misinformation detection is decentralized, secure, verifiable, and contextually aware. It also discusses the architectural, infrastructural, and operational implications of creating such interoperable systems. These systems are required to be designed carefully keeping in mind the risks and regulatory implications that are inherent in these innovations (Parvez et al., 2018).

### 13.4.1 Blockchain for content authentication and provenance

Blockchain technology provides its great support in misinformation detection and prevention tasks. With its decentralized and tamper-resistant structure, it holds revolutionary power in making digital content more transparent and trustworthy (Casino, Dasaklis, and Patsakis, 2019).

By recording content metadata like timestamps, authorship, source URLs, and edit history upon a public or permission blockchain, these platforms are helpful in identifying misinformation. They can verify if a piece of content has been tampered with or falsified. Smart contracts can be used to automate verification procedures in detecting misinformation, and cryptographic hashes can be used to compare published content against the original content to identify unauthorized changes (Christidis and Devetsikiotis, 2016).

This integration is especially beneficial for fact-checking initiatives, news providers, and social networks for misinformation detection. It is very helpful by allowing real-time verification at scale. When coupled with AI-based detection tools, blockchain can offer an audit trail to manage misinformation. It helps determine content credibility and curb the spread of disinformation, particularly in high-stakes domains like elections or public health emergencies. Figure 13.2 shows that the watermarks, metadata etc. are recorded on a blockchain, ensuring an immutable record (Galason, 2023).

In addition to safeguarding metadata, digital identity systems such as digital signatures supported by blockchain are being used. They ensure that only institutions or approved content authors are allowed to share high-impact or sensitive information to maintain secrecy. By linking digital content to verified decentralized identifiers (DIDs), end-users can link the content back to a responsible originator and secure their content. This shuns anonymous or automated sharing of falsehood and manages misinformation. Moreover, it establishes a web of trust among end-users, platforms, and content authors. Such an identity-attributing parameter is particularly

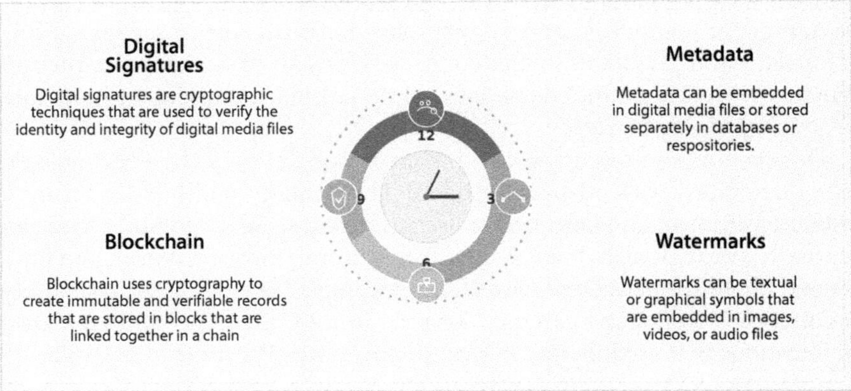


*Figure 13.2* Block chain registration.

beneficial in times of crisis, where misinformation spreads fast and authoritative content must be protected instantly.

Second, blockchain enables decentralized content moderation sites to manage misinformation. It encourages openness and democratic governance, which results in secure and safe outcomes. Traditional content moderation basically takes place in a black-box system of centralized platforms with lack of transparency. However, blockchain-enabled governance provides transparent results. It involves stakeholders (reporters, researchers, fact-checkers, and users) in voting or consensus processes to verify the content. It allows them to validate the legitimacy of content and ensures enhanced protection. Decentralized moderation of this kind encourages accountability and neutrality. It also provides its support to suppress the censorship risks. Combined with AI moderation and cryptographic validation, blockchain offers an enhanced defense against disinformation. This convergence of technologies results in improving technical robustness and public trust.

### 13.4.2 Digital identity and AI-based credibility scoring

AI systems are very helpful in authenticating digital identities and providing credibility scores. These systems can assess behavioral signals, communication patterns, and history content. They are capable of delivering users a credibility rating based on the content (Naik and Jenkins, 2019). When implemented together with DIDs or biometric authentication systems, AI systems provide better authentication and safety. They authenticate identities without sharing any type of sensitive personal information and provide increased protection (Jain and Kumar, 2020). AI-based credibility scoring is also helpful for separating genuine players from maligned ones.

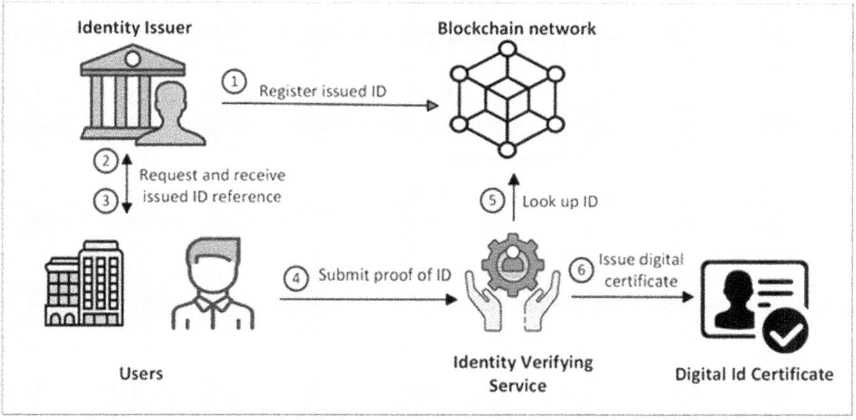


*Figure 13.3* Block chain-based digital identity authentication architecture.

Embedding such mechanisms within sharing platforms can provide users with context-specific notifications to safeguard them. This is not centralizing the user data, thus maintaining the safety. Moreover, it is adhering to the privacy-conserving principles, especially if complemented with federated identity patterns, as shown in Figure 13.3 (Gebreab et al., 2022). Within a federated identity system, privacy-preserving AI-driven credibility scoring agents play an important role. They inspect on-chain attested credentials and off-chain behavioral metadata for the verification purpose. They create live trust indicators that can be directly integrated into sharing platforms to protect the content. When a user is exposed to content, the system cross-checks the post's decentralized ID for the safety point of view. It then dynamically marks low-credibility or unverifiable sources (without centralizing individual data). This is done by utilizing the federated consensus among identity issuers and verifiers, thus providing safety and security. This system provides context-aware notifications that respect user privacy while ensuring strong authenticity.

## 13.5 EXPLAINABILITY, ACCOUNTABILITY, AND TRUST IN FUTURE SYSTEMS

Explainability, accountability, and trust are very important factors that require strong focus. The more deeply generative AI models embed into detection and misinformation mitigation, the more these factors become imperative. It becomes essential that transparency, governance, and trust be emphasized. Technical performance is also an important factor, but technology's credibility and prosperity ultimately depend on many other factors. These factors include how and why decisions are made, who is held accountable, and if the values of society are being taken into consideration. In this context, we explain how explainability, accountability, and trust will be integral to next-generation AI-based misinformation systems. These systems will not be black-boxes anymore, unlike traditional systems. This transformation in misinformation systems will provide outcomes with enhanced accuracy. Thus, these systems not only allow efficiency but also transparency, impartiality, and dependability.

### 13.5.1 Explainable AI (XAI) for misinformation detection

As AI models become increasingly sophisticated, it becomes difficult to understand how they reason. It is particularly a problem associated with the general deployment of LLMs. This lack of transparency is a huge impediment in high-risk applications such as detecting misinformation, where transparency and trust are essential (Doshi-Velez and Kim, 2017).

In misinformation, XAI plays an important role: it serves two-fold function. It assists system developers, researchers, and policymakers in the auditing process. It guides them on how an AI system detects falsity, manipulative content, or propaganda (Guidotti et al., 2018). It also helps end-users by providing clear explanations regarding why content was indicated. This type of transparency is thereby fostering trust and reducing feelings of censorship or algorithmic prejudice.

Various approaches are being designed to detect and prevent misinformation. These are as given below:

- **Attention mechanism visualization:** It consists of the transformer-based architecture to manage misinformation. The examples include GPT and BERT models which are being used significantly. These models use attention layers to weight input tokens. Visualizing these weights allows analysts to monitor the process. The analysts can see what part of the input the model was attending to, when it made a classification or when the decisions were made.
- **Post-hoc explanation methods:** *Posthoc* explanation methods include SHapley Additive exPlanations (SHAP) and Local Interpretable Model-agnostic Explanations (LIME) models. These models are employed to produce feature-level attributions of predictions to manage misinformation. They can, for example, indicate which phrases played the most significant roles in leading the model to classify a piece of text as deceptive (Lundberg and Lee, 2017).
- **Counterfactual reasoning:** Here, researchers can infer what features are most important in making decisions by making slight modifications in inputs and observing the impact on model predictions. This helps in uncovering the excessive reliance on stylistic features (e.g., sensationalism) rather than factual inaccuracies (Mittelstadt et al., 2016).

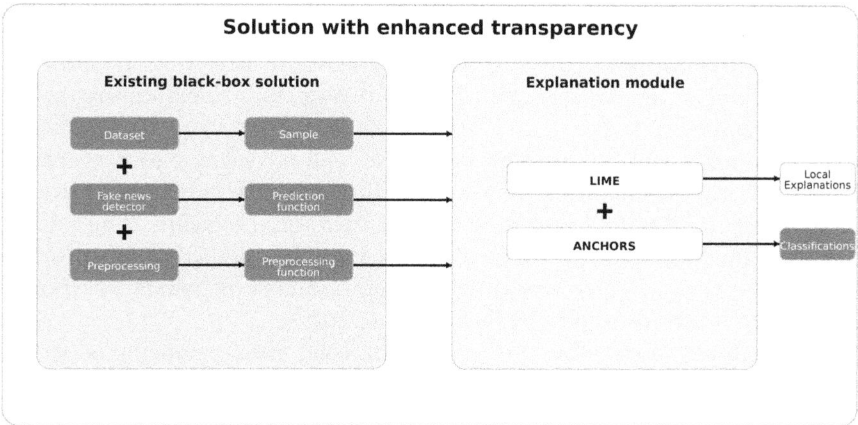


*Figure 13.4* Explainability method for the BERT-based model in fake news detection.

- **Natural language rationales:** It includes training models to return human-interpretable reasons for their output. As an example, instead of directly tagging a post, the system can say, "This claim has been debunked by several verified fact-checking sources."

Explainability is a concern of technical, legal, and ethical aspects. AI policies worldwide, including the European Union's AI Act, present that transparency and explainability are primary issues of regulatory compliance, especially in high-risk AI systems. Explainability may also be used to deter attacks on misinformation-detection systems pertaining to adversarial manipulations, spurious correlations, or inherent biases.

As information consumption is increasingly mediated by AI, being able to read and understand how it identifies misinformation would be a possibility for democratic debates and for erecting public trust. Hopefully, further evolution of XAI will combine deep technical advances with the design needs occurring from user requirements, such that the transparency becomes transparent to both the experts and laymen. One such approach is presented in Figure 13.4 (Szczepański et al., 2021). It introduces a new explanation technique for BERT-based fake news detection. It is based on mixing LIME and anchors for unleashing the decision process of the model with minimal modifications to the current setup.

### 13.5.2 Accountability mechanisms in AI-powered misinformation systems

The usage of AI systems in spotting and combating misinformation is growing nowadays. It raises the need for sound accountability frameworks for better management. Accountability is the main factor that requires strong attention. This factor is attributed to the creation, deployment, and outcomes of AI-based systems. It is, therefore, required to employ well-structured accountability frameworks with the usage of AI in detecting misinformation. Otherwise, it may result in technical failures, ethical controversies, and legal issues.

Accountability in AI systems should be well structured to prevent the failures. It involves defining the responsibilities at each step of the methodology lifecycle of the AI system. It is done through the means of successful deployment, surveillance, and user engagement. For detecting misinformation, developers who are revealing the data streams on which the models are trained are responsible (Veale and Bins, 2017).

Transparency also plays an important role, thus requiring a strong emphasis. It serves as one of the fundamental building blocks of accountability. The study by Wieringa illustrates that AI systems which are being used to detect and prevent misinformation must be separately audited and evaluated (Wieringa, 2020). These evaluations can be helpful in identifying latent effects (biases being reinforced and opinions being suppressed). The regular audits are required to examine fairness, explainability, and robustness for successful implementation. They also check that compliances are aligned with ethical standards or regulatory terms to maintain the safety and security. Publicization of audit results strengthens the institutional legitimacy and thus enhances the trust. It also provides a venue for corrective measures to streamline the functions.

Furthermore, there is a need for recourse and redress mechanisms to protect the individuals. Users, for example, whose content is removed must be given a chance to understand the reasons. Moreover, they can appeal for their rights through easy appeal mechanisms. This becomes particularly important when misinformation detection systems are scaled and impact political discourse. Accountability, therefore, is served as a two-fold function: the responsibility to make the correct choice as well as correct the mistakes.

Traceability is also one of the central features of accountability. Traceability of data and decisions is crucial, which needs to be taken care of. AI systems used for misinformation detection need to be thoroughly documented about how decision-making was performed for better understanding. It is also required to take care of other factors such as what data went into it and the amount of human intervention involved. Traceability enables correct auditing and thus enhances the outcomes. It supports explainability efforts and enables learning from past errors, allowing timely implementation of corrective measures. Traceable systems are also able to respond more reactively in detecting misinformation and therefore provide results with better accuracy.

Regulatory compliances are also a very essential part of accountability. They will become the formal dimension of accountability as per the European Union's AI Act (Cath, 2018). Such laws provide risk-based categories for AI systems so that the measures can be taken accordingly. They assign legal roles to developers, deployers, and platform operators for successful implementation. In the coming years, failing to make accountability operational will be harmful and might result in reputation loss. It may also bring legal consequences, particularly in high-stakes areas like information integrity and electoral security (Goodman and Flaxman, 2017).

Accountability in AI-enabled misinformation systems must be systemic and backed by legal, organizational, and societal infrastructures. As techniques for spreading misinformation are growing, there is the need for accountability systems. These systems will support to govern the use of AI to counter the misinformation. For successful implementation, the collaboration between technologists, regulators, civil society groups, and the public is required. This collaboration ensures that the use of AI in information environments is not only effective but also fair, transparent, and democratic. Figure 13.5 from the OECD outlines a comprehensive AI accountability framework. This framework emphasizes the shared responsibilities among developers, deployers, and integrators (OECD, 2023). It highlights practices such as risk assessments, documentation, and auditability, which are crucial for ensuring transparency and accountability in AI systems. Monitoring, reviewing and governance further add accountability.

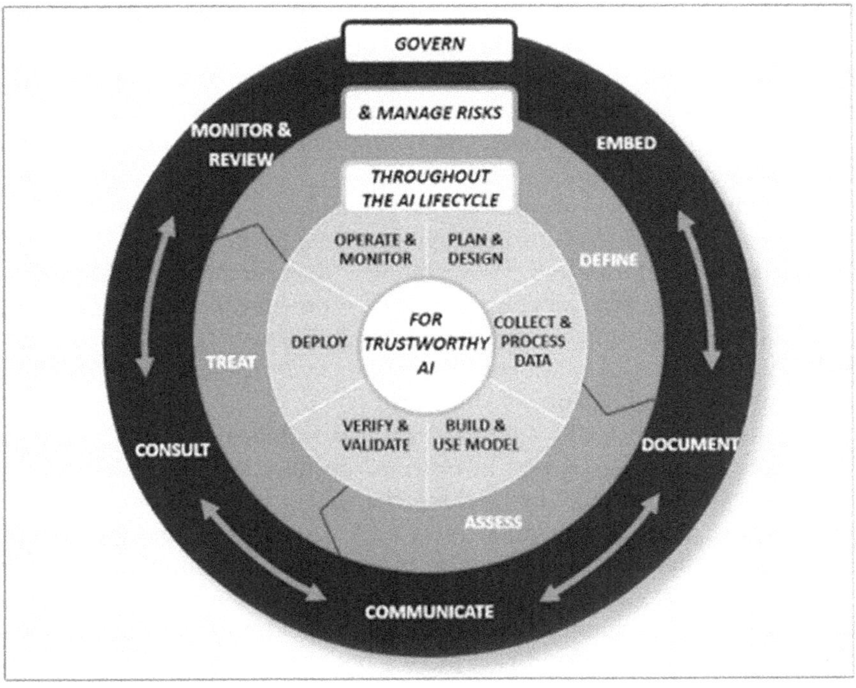


*Figure 13.5* OECD AI accountability framework.

## 13.6 PROACTIVE AND PREDICTIVE MISINFORMATION MANAGEMENT

Even as war against misinformation continues, as the scales of the struggle expand bigger and become more complex day by day, it has become apparent that conventional, solely reactive approaches have their limitations. Before misinformation could be found and addressed, it has already caused immense harm in eroding public trust, distorting democratic debate, or posing danger to public health. AI-based systems utilized for spreading misinformation in the future will evolve from detection to proactive and predictive systems that peer into the future to foretell threats, act early, and reduce the spread before misinformation gets entrenched.

Active misinformation management involves using AI systems that will constantly monitor and scan new patterns of content, social media discourse, and internet ecosystems, which will assist in identifying potentially harmful narratives early on. These systems use a blend of real-time data streams, NLP, and knowledge graphs to watch for anomalies and trending topics that are far from established truths or exhibit attributes proven to be connected with misinformation. Most importantly, these systems are designed to alert dangerous content ahead of time for increased scrutiny, fact-checking, or situational framing.

Predictive models are very helpful and are being used frequently to forecast misinformation incidents before they happen. Based on past trends, behavior, and patterns, AI can predict when and where misinformation is most likely to happen. The predictive power of AI-based models can facilitate strategic preparedness. It enables stakeholders to pre-position counter-narratives and strengthen public training and education programs.

The study by Kim et al. illustrated that network analysis is one of the major components of predictive misinformation preemption (Kim et al., 2020). Analyzing how information flows among influencer networks allows AI systems to identify pre-campaign potential super-spreaders. This opens up possibilities for pre-emptive actions against misinformation. These include early interaction with platform moderators and reporting fact-checking organizations ahead of time.

Emotion and sentiment analyses are also significant. Misinformation campaigns tend to be created in order to produce strong emotions such as fear, anger, or outrage. By tracking changes in the emotions and sentiments of the general population on a scale, AI systems are able to find outlier spikes that are associated with an attempt at a manipulative content strategy (Shu et al., 2020). Real-time responses to such notifications can be used to modulate content recommendation algorithms, remove polarizing dialogues, or give more credible voices a boost to put some balance back into discussion.

Synthetic media prediction is another new methodology. Due to the growing proficiency of generative models in producing highly realistic deep fakes and text-based fakes, AI systems have to guess which type of media, stories, or faces stand the best chance of being manipulated in the next attacks on truth. This can be used to develop tools in advance, e.g., detectors of new image generation methods or new content moderation policies that are sensitive to new generative AI threats. Ethical regulation needs to be matched with predictive and proactive systems, which are of particular interest. Disinformation prediction would require a certain amount of speculation, which would lead to false positives or overreach (Jobin et al., 2019). The danger of such systems identifying content erroneously has not yet caused harm and should be handled by a delicate balance, human intervention, and communication of the predictive capabilities. Freedom of expression and avoiding the development of such systems as a way of mistrust/misinformation should be provided (not being regarded as too interventionist or prescriptive) (Narayanan et al., 2008).

Collaboration will play a central role in the process of making the management proactive and predictive misinformation effective as no platform or organization has the big picture. Data exchange among platforms, governments, research organizations, and civil society will result in a more cohesive picture about threats that are emerging. This collaboration can be facilitated by federated learning and privacy-preserving AI techniques, and thus user anonymity and data safety is guaranteed.

The future of misinformation defense is based on the quality of our ability to detect lies after they have been posted. It also depends on the quality to predict them, recognize them, and shut them off before they disseminate. AI will be the core of this change as a device to tell the truth. It also acts as a supporter of the spread of the truth and trust in the digital-first world. Figure 13.6 presents the automatic fake news detection system (Al-Obaidi and Çağlıkantar, 2024). It assembles various components such as data collection, preprocessing, feature extraction, and classification to observe and curb misinformation successfully. The proposed automatic fake news detection system workflow starts with the collection of news datasets. It then works with preprocessing steps of cleaning and tokenizing to make it readable by the model. This processed text is then input into deep learning networks such as LSTM, CNN-LSTM, CNN-GRU, and LSTM Auto Encoder. These models are trained to learn language and contextual features that separate fake and real news. The performance of each model is measured on benchmark datasets. The performance is mostly quantified in terms of accuracy. According to the results, the system finds the most appropriate model to fit a particular dataset. It shows that the effectiveness of different systems is based on the qualities of the dataset instead of only a universally better model.

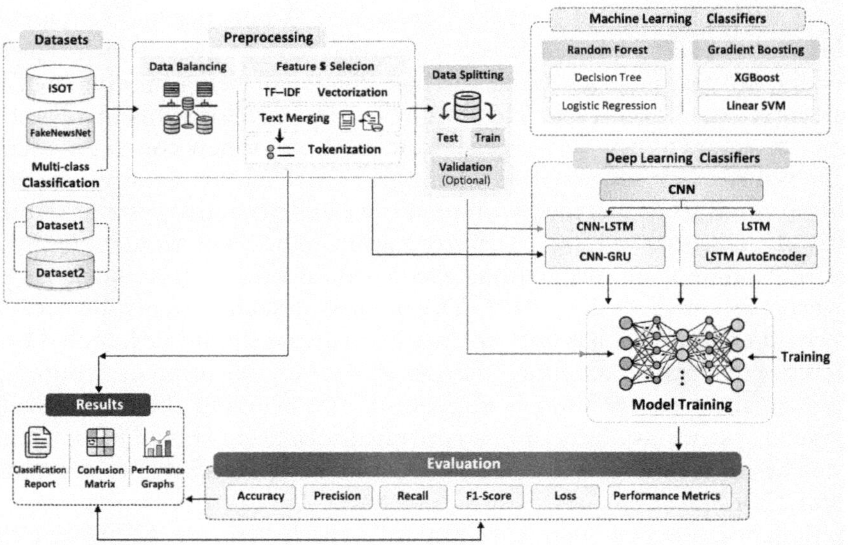


*Figure 13.6* Architecture of the automatic fake news detection system.

## 13.7 ETHICAL, SOCIAL, AND LEGAL CONSIDERATIONS

AI technology is becoming central to identification and prevention of misinformation. The development and use of AI-based systems raise profound ethical, social, and legal concerns. Such concerns not only impact the technical performance but also governance, public trust, and democratic integrity. Emerging systems tend to be very powerful and anti-falsehood countermeasures. They must operate in a fair, inclusive, and legally sensible way to fight against misinformation.

Morally, the greatest concern is how to reconcile freedom of expression with controlling noxious material. AI applications for censoring or labeling information risk encroaching upon individuals' rights of communication and exposure to a variety of viewpoints (Cath et al., 2018). It has been found that misinformation has disastrous effects; overly intrusive or opaque interventions can potentially cause alarming effects or disproportionately silence minority voices, so there is a need to design AI systems with the ethics of proportionality, non-discrimination, and explainability to curb the misinformation. These systems will enable users to observe how decisions are made and allow them to challenge where necessary.

Another major ethical concern which needs to be taken into consideration is the possibility of algorithmic bias. The study by Mittelstadt et al. states that misinformation detection systems will only be as unbiased as the data they are trained on (Mittelstadt et al., 2016). If the training data reflect societal biases or historical prejudices, AI systems will replicate and even exaggerate them by misclassifying dialects, discriminating against particular groups, or penalizing nonconventional discourse. Developers must actively audit, and minimize, biases in datasets as well as in model behavior. This can include involving diverse communities in the design process and ensuring that the minority opinions are not excluded or misrepresented. Socially, the use of AI to flag or filter content can have the effect of subtly but noticeably altering public discourse. As users become aware that AI systems are monitoring or policing content, their communication practices may not always change for the better.

There is a risk of self-censorship, polarization, and loss of trust in platforms if users perceive AI decisions as biased or ideologically based (Eubanks, 2018). So social acceptability is no less important than technical performance. Trust with users needs to be built by being transparent, having human oversight, and maintaining constant public debate on how and why AI is being employed. Power concentration is a more general concern. Large tech firms, governments, and platform providers exert a great deal of control over how AI moderation tools are constructed and implemented. These dangers result in asymmetries of power where it is the privilege of a limited set of actors to decide what is "truth" or "misinformation" (Yeung, 2018).

Prevention lies in making future regimes go through inclusive processes, thus maintaining transparency and offering better control. It includes the involvement of a civil society, the academic community, independent media, and international organizations. Decentralized governance must be established to ensure informational integrity in an open society. It has been observed that legally, the misinformation detection has a connection with data protection, content moderation, and liability law. Legislative frameworks such as the European Union's AI Act and General Data Protection Regulation (GDPR) begin to establish legal boundaries on AI-based moderation to maintain safety and security. These rules and policies require openness, human oversight, and accountability for high-risk systems. Future AI systems will be required to comply with such laws for upholding user rights and democratic principles.

Additionally, there is a need to establish laws which will dictate how tools for detecting misinformation can function globally. New legislations will likely introduce risk categorization and audit requirements to AI systems that influence public dialogue, as stated by Kroll et al. (2017). Developers must expect increased documentation duties, third-party testing, and user recourse processes. As the regulation comes into view, it is important to follow the compliances and use AI responsibly. The ethical and legal obstacles to identifying misinformation must be viewed as dynamic rather than static. As social norms, legal frameworks, and disinformation tactics are constantly evolving, ethical and legal rails will have to be adaptive, permitting revision and participatory reform (Gasser and Almeida, 2017). Future systems must be developed considering technical, ethical, and legal feasibilities. In summary, the fight against misinformation cannot be waged by algorithms only; it requires other solutions as well (Wachter et al., 2017). There is a need to consider ethics, laws, and public values while designing AI systems. Making these systems protect basic rights and safeguard trust is perhaps one of the most important challenges in the age of smart information technologies (Goodman & Flaxman, 2017). As shown in Figure 13.7, the Venn diagram presents an overlap of important ethical principles in AI development (Radanliev, 2025). It indicates that transparency, fairness, explainability, accountability, and trust are very important factors that need to be considered while developing and using AI systems. It further indicates that equity, data protection, and responsible use of AI will be the demand of future systems.

In order to overcome the ethical challenges, some approaches need to be taken into account. First, AI moderation can be used together with

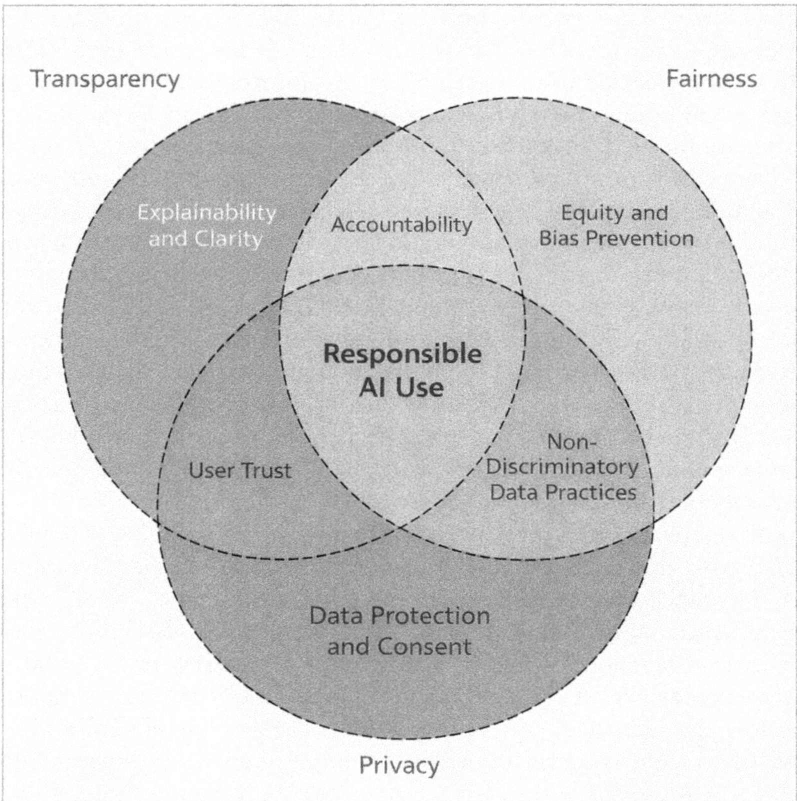


*Figure 13.7* Ethical considerations in AI development.

human-in-the-loop systems. Here, flagged content is reviewed by human services, especially in the case of subtle or controversial content (Udupa, 2021). Second, AI systems need to be transparent and explainable to provide moderation decisions. These may be both audited and comprehensible to ensure that the results reflect any latent bias. Third, we need nonbiased and varied datasets to reduce systemic bias. It is better to include marginalized voices so that AI does not make dominant perspectives the default truth (Nahmias and Perel, 2021). Lastly, consistent ethical impact assessment projects should be implemented to fight against misinformation. These projects provide their support to determine the effects of moderation practices. These methods help enable organizations to optimize policies and algorithms in the long run (Banchio, 2024). All the mentioned measures would help enhance the ethical practices of AI-driven misinformation control.

## 13.8 CHALLENGES AND OPEN RESEARCH QUESTIONS

Though AI-powered systems are very helpful in detecting and preventing misinformation, many challenges still exist. One of the key technical challenges is models' susceptibility to adversarial attacks (Carlini et al., 2019). It harms the safety and security components. Such insidious manipulations are leading robustness to remain a constant issue.

As per the study by Zhang et al, it is difficult to provide performance in languages, cultures, and modalities, so it becomes a big challenge while implementing the solutions (Zhang et al., 2019). Numerous AI models are highly biased toward high-resource languages; this bias distracts the actual outcomes. A significant gap is found in the capability of detecting misinformation for underrepresented linguistic and regional groups. Also, it is still challenging to detect multimodal misinformation which includes pictures, videos, and audio in a single framework.

Data shortages and quality are also important issues which impact the implementation of AI-based systems. The study by Sharma et al. states that labeled misinformation sets are few and tend to fall behind new narratives (Sharma et al., 2020). This impedes supervised learning methods and introduces response time lag. Moreover, model relevance over time necessitates processes for continuous learning, which is still an active area of research.

From an ethical and operational perspective, the lack of explainability is also a big concern in most of the AI systems. The users do not know why and how decisions are being made. It restricts user trust, transparency, and regulatory adherence. Most of the users cannot discern why certain content is filtered or not. It creates issues related to bias, censorship, and accountability as per the research by Jiang et al. (2019).

Scalability is another significant challenge. Implementing sophisticated AI solutions in real-time poses scalability issues. This problem is particularly associated with the implementation of edge devices and federated systems. It raises the need to optimize the models to perform under resource constraints like limited computing capability, bandwidth, and sporadic connectivity (Yao et al., 2019).

Finally, open research questions exist. These questions include how to predict and anticipate misinformation without harming the freedom of speech. It has been stated by Doshi and Kim in one of their studies that it is difficult to maintain democratic values in the presence of proactive censorship (Doshi-Velez and Kim, 2017). It is also difficult to examine the social effects of algorithmic moderation. It includes its impact on marginalized communities. Certain questions related to legal issues regarding approvals and adoptions also pose challenges in the implementation of AI systems.

It has been found that despite technological advancements against misinformation, there exist significant challenges which need to be addressed yet. Addressing these challenges will require further investigations in AI research, ethics, and law (Roberts, Stewart, and Tingley, 2018).

## 13.9 CONCLUSION

The misinformation complexity is growing in the current age of digital information. AI introduces unparalleled opportunities to fight against misinformation. This chapter has covered AI technologies from LLMs and multimodal architectures to federated learning and XAI. The growing capabilities of these technologies in enabling smarter, faster, and more context-aware detection and prevention of misinformation have been explained. The future of misinformation management lies in an advanced combination of innovation, governance, ethics, and social responsibility. The successful implementation of AI demands transparency, fairness, and accountability. It has been observed that the combination of different technologies like blockchain, edge computing, and digital identity systems further enhanced the outcomes. They resulted in building trust, decentralization, and responsiveness in AI systems. But despite much advancement, the challenges still exist. From biasing and data constraints to scalability and legality, the path ahead is filled with technical and ethical challenges. Predictive and proactive AI systems further provided early detection and management. They must be designed on the basis of interdisciplinary cooperation and participative governance. Finally, the destiny of misinformation detection will rest on the quality of our ability to detect lies, stay alert, and by fair AI innovations. With responsible development and usage, AI can become a foundation for maintaining the truth and trust. This chapter will help academicians, developers, and researchers to investigate further in the field and make more developments.

## REFERENCES

Alayrac, J.-B., Donahue, J., Luc, P., et al. 2022. Flamingo: A visual language model for few-shot learning. arXiv Preprint arXiv:2204.14198.

Al-Obaidi, S. A., and Çağlıkantar, T. 2024. Automated fake news detection system. *Iraqi Journal for Computer Science and Mathematics,* 5(4). https://doi.org/10.52866/2788-7421.1200

Banchio, P. R. 2024. Legal, ethical and practical challenges of AI-driven content moderation. *SSRN Electronic Journal*, 1–4.

Bender, E. M., Gebru, T., McMillan-Major, A., and Shmitchell, S. 2021. On the dangers of stochastic parrots: Can language models be too big? In *Proceedings of the 2021 ACM Conference on Fairness, Accountability, and Transparency* (pp. 610–623). Virtual (online) event associated with Toronto, Canada from March 3–10, 2021.

Brown, T., Mann, B., Ryder, N., Subbiah, M., Kaplan, J. D., Dhariwal, P., and Amodei, D. 2020. Language models are few-shot learners. *Advances in Neural Information Processing Systems*, 33, 1877–1901.

Brundage, M., Avin, S., Wang, J., et al. 2020. Toward trustworthy AI development: Mechanisms for supporting verifiable claims. arXiv Preprint arXiv:2004.07213.

Carlini, N., Athalye, A., Papernot, N., Brendel, W., Rauber, J., Tsipras, D., Goodfellow, I. J., and Madry, A., 2019. Evaluating and understanding the robustness of adversarial training. In *Proceedings of the 36th International Conference on Machine Learning* (Vol. 97, pp. 2532–2541). Long Beach, CA, USA.

Casino, F., Dasaklis, T. K., and Patsakis, C. 2019. A systematic literature review of blockchain-based applications: Current status, classification and open issues. *Telecommunications Systems*, 71, 339–358.

Cath, C.2018. Governing artificial intelligence: Ethical, legal and technical opportunities and challenges. *Philosophical Transactions of the Royal Society A*, 376(2133), 20180080.

Cath, C., Wachter, S., Mittelstadt, B., Taddeo, M., and Floridi, L. 2018. Artificial intelligence and the "good society": The US, EU, and UK approach. *Science and Engineering Ethics*, 24(2), 505–528.

Christidis, K., and Devetsikiotis, M. 2016. Blockchains and smart contracts for the internet of things. *IEEE Access*, 4, 2292–2303.

Doshi-Velez, F., and Kim, B. 2017. Towards a rigorous science of interpretable machine learning. arXiv Preprint arXiv:1702.08608.

Eubanks, V. 2018. *Automating Inequality: How High-tech Tools Profile, Police, and Punish the Poor*. St. Martin's Press.

Galason, E. 2023. Digital authenticity: Provenance and verification in AI-generated media. Numbers Protocol. https://numbersprotocol.io/blog/digital-authenticity-provenance-and-verification-in-ai-generated-media.

Gasser, U., and Almeida, V. A. 2017. A layered model for AI governance. *IEEE Internet Computing*, 21(6), 58–62.

Gebreab, S. A., Hasan, H. R., Salah, K., and Jayaraman, R. 2022. NFT-based traceability and ownership management of medical devices. *IEEE Access*, 10, 126394–126411.

Goodman, B., and Flaxman, S. 2017. European Union regulations on algorithmic decision-making and a right to explanation. *AI Magazine*, 38(3), 50–57.

Grill, J.-B., Strub, F., Altché, F., et al. 2020. Bootstrap your own latent: A new approach to self-supervised learning. *Advances in Neural Information Processing Systems*, 33, 21271–21284.

Guidotti, R., Monreale, A., Ruggieri, S., Turini, F., Giannotti, F., and Pedreschi, D. 2018. A survey of methods for explaining black box models. *ACM Computing Surveys*, 51(5), 1–42.

Jain, A., and Kumar, M. 2020. AI-enabled user credibility scoring for social media platforms. In *2020 IEEE International Conference on Big Data* (pp. 5264–5273). Atlanta, Georgia: IEEE. (As a virtual/online conference due to the COVID-19 pandemic).

Jain, S., and Wallace, B. C. 2019. Attention is not explanation. In *Proceedings of the 2019 Conference of the North American Chapter of the Association for Computational Linguistics: Human Language Technologies* (pp. 3543–3556). Minneapolis, Minnesota: NAACL-HLT 2019.

Jiang, L., et al. 2019. Improving robustness of deep neural networks via stability training. In *Proceedings of the IEEE Conference on Computer Vision and Pattern Recognition* (pp. 4480–4488).

Jobin, A., Ienca, M., and Vayena, E. 2019. The global landscape of AI ethics guidelines. *Nature Machine Intelligence*, 1, 389–399.

Kairouz, P., McMahan, H. B., Avent, B. et al. 2021. Advances and open problems in federated learning. *Foundations and Trends in Machine Learning*, 14(1–2), 1–210.

Kim, H., Shah, N., and Liu, H. 2020. Deep learning for misinformation detection on social media: A survey and new perspectives. *ACM Transactions on Information Systems*, 38(4), 1–37.

Kroll, J. A., Huey, J., Barocas, S. et al. 2017. Accountable algorithms. *University of Pennsylvania Law Review*, 165(3), 633–705.

Lundberg, S. M., and Lee, S.-I. 2017. A unified approach to interpreting model predictions. *Advances in Neural Information Processing Systems*, 30, 4768–4777.

Maras, M.-H., and Alexandrou, A. 2019. Determining authenticity of video evidence in the age of artificial intelligence and in the wake of deepfake videos. *International Journal of Evidence & Proof*, 23(3), 255–262.

Mittelstadt, B. D., Allo, P., Taddeo, M., Wachter, S., and Floridi, L. 2016. The ethics of algorithms: Mapping the debate. *Big Data & Society*, 3(2), 2053951716679679.

Morkunas, V. J., Paschen, J., and Boon, E. 2019. How blockchain technologies impact your business model. *Business Horizons*, 62(3), 295–306.

Nahmias, Y., and Perel, M. 2021. The oversight of content moderation by AI: Impact assessments and their limitations. *Harvard Journal on Legislation,* 58(1), 145–194.

Naik, N., and Jenkins, P. 2019. Ubiquitous digital identity for a decentralized world. *Journal of Computer Networks and Communications*, 2019, Article 6704159.

Narayanan, A., Hu, Y., and Shmatikov, V. 2008. De-anonymizing social networks. In *2008 IEEE Symposium on Security and Privacy* (pp. 173–187). Oakland, CA: IEEE.

Nguyen, T. T., Pathirana, P. N., Nguyen, D. C., Seneviratne, A., and Li, J. 2021. Federated learning for smart healthcare: A survey. *ACM Computing Surveys*, 55(3), 1–37.

OpenAI.2023. GPT-4 technical report. arXiv Preprint arXiv:2303.08774.

Organisation for Economic Co-operation and Development. 2023. Advancing accountability in AI.

Parvez, I., Rahmati, A., Guvenc, I., Sarwat, A. I., and Dai, H. 2018. A survey on low latency towards 5G: RAN, core network and caching solutions. *IEEE Communications Surveys & Tutorials*, 20(4), 3098–3130.

Radanliev, P. 2025. AI ethics: Integrating transparency, fairness, and privacy in AI development. *Applied Artificial Intelligence*, 39(1), 1–41.

Ribeiro, M. T., Singh, S., and Guestrin, C. 2016. "Why should I trust you?" Explaining the predictions of any classifier. In *Proceedings of the 22nd ACM SIGKDD International Conference on Knowledge Discovery and Data Mining* (pp. 1135–1144). San Francisco, Ca, USA.

Roberts, M. E., Stewart, B. M., and Tingley, D. 2018. Navigating the AI ethics landscape: Challenges and opportunities. *Science*, 361(6400), 750–752.

Satyanarayanan, M. 2017. The emergence of edge computing. *Computer*, 50(1), 30–39.

Shahbazi, Z., and Byun, Y. C. 2021. Fake media detection based on natural language processing and blockchain approaches. *IEEE Access*, 9, 128442–128453.

Sharma, U., Saran, S., and Patil, S. M. 2020. Fakenews detection using machine learning algorithms. *International Journal of Engineering Research & Technology (IJERT)*, 9(03), 1–10.

Shi, W., Cao, J., Zhang, Q., Li, Y., and Xu, L. 2016. Edge computing: Vision and challenges. *IEEE Internet of Things Journal*, 3(5), 637–646.

Shu, K., Sliva, A., Wang, S., Tang, J., and Liu, H. 2020. Fake news detection on social media: A data mining perspective. *ACM SIGKDD Explorations Newsletter*, 19(1), 22–36.

Szczepański, M., Pawlicki, M., Kozik, R., and Choraś, M. 2021. New explainability method for BERT-based model in fake news detection. *Scientific Reports*, 11(1), 23705.

Thorne, J., and Vlachos, A. 2018. Automated fact checking: Task formulations, methods and future directions. In *Proceedings of the 27th International Conference on Computational Linguistics* (pp. 3346–3359). Santa Fe, New Mexico, USA.

Tobarra, L., Ros, S., Robles-Gomez, A., Hernandez, R., Caminero, A. C., and Pastor, R. 2019. An authentication system based on digital identity using blockchain. *IEEE Access*, 7, 146595–146604.

Udupa, S. 2021. Digital technology and extreme speech: Approaches to counter online hate. Commissioned research paper for the United Nations Peacekeeping Technology Strategy.

Veale, M., and Binns, R. 2017. Fairer machine learning in the real world: Mitigating discrimination without collecting sensitive data. *Big Data & Society*, 4(2), 201712.

Wachter, S., Mittelstadt, B., and Floridi, L. 2017. Why a right to explanation of automated decision-making does not exist in the General Data Protection Regulation. *International Data Privacy Law*, 7(2), 76–99.

Weller, A. 2019. Transparency: Motivations and challenges. In Samek, W., Montavon, G., Vedaldi, A., Hansen, L. K., Müller, K. -R. (eds), *Explainable AI: Interpreting, Explaining and Visualizing Deep Learning* (pp. 23–40). Springer.

Whittaker, M., Crawford, K., Dobbe, R., et al. 2018. AI Now report 2018. AI Now Institute, New York University.

Wieringa, M. 2020. What to account for when accounting for algorithms: A systematic literature review on algorithmic accountability. In *Proceedings of the 2020 Conference on Fairness, Accountability, and Transparency* (pp. 1–18). Barcelona, Spain.

Yao, Q., et al. 2019. Federated learning with non-IID data. arXiv Preprint arXiv:1906.08935.

Yeung, K. 2018. Algorithmic regulation: A critical interrogation. *Regulation & Governance*, 12(4), 505–523.

Zheng, S., Song, Y., Leung, T., and Goodfellow, I. J. 2016. Improving the robustness of deep neural networks via stability training. In *Proceedings of the IEEE conference on computer vision and pattern recognition* (pp. 4480–4488). Las Vegas, NV, USA. IEEE.

Zhao, Y., Li, M., Lai, L., Suda, N., Civin, D., and Chandra, V. 2018. Federated learning with non-IID data. *arXiv* Preprint arXiv:1806.00582

Zhang, C., Gupta, A., Kauten, C., Deokar, A. V., and Qin, X. 2019. Detecting fake news for reducing misinformation risks using analytics approaches. *European Journal of Operational Research,* 279(3), 1036–1052. https://doi.org/10.1016/j.ejor.2019.06.022.

Zhou, X., and Zafarani, R. 2018. A survey of fake news: Fundamental theories, detection methods, and opportunities. *ACM Computing Surveys*, 53(5), 1–40.

# Index

For Product Safety Concerns and Information please contact our EU
representative GPSR@taylorandfrancis.com
Taylor & Francis Verlag GmbH, Kaufingerstraße 24, 80331 München, Germany

www.ingramcontent.com/pod-product-compliance
Ingram Content Group UK Ltd.
Pitfield, Milton Keynes, MK11 3LW, UK
UKHW022219190926
475702UK00003B/445

* 9 7 8 1 0 4 1 0 4 0 4 0 8 *